THE SCIENCE EXPLORER

I0483439

VARDHAN GOYAL

XpressPublishing
An imprint of Notion Press

XpressPublishing
An imprint of Notion Press

No.8, 3rd Cross Street,CIT Colony,
Mylapore, Chennai, Tamil Nadu-600004

ISBN 978-1-64783-732-7

"Two things are infinite: the universe and human stupidity; and I'm not sure about the universe."
—Albert Einstein

Contents

Acknowledgements

First and foremost, praises and thanks to the God, the Almighty, for His showers of blessings throughout my research work to complete the research successfully.

I cannot express enough thanks to the Members of the Editorial Board for their continued support and encouragement. I offer my sincere appreciation for the learning opportunities provided by them.

My completion of this project could not have been accomplished without the support of my classmates, Ansh, Vansh, Akhilesh, Eshan, Gitanjay, Yashwant, and all my friends whom I have missed out. Thank you for allowing me time away from you to research and write. You deserve a trip to Disney! I am extremely grateful to my sister and my parents for their love, prayers, caring and sacrifices for educating and preparing me for my future. I would like to express my deep and sincere gratitude to my research supervisor, Mr. J.D. Nanda for giving me the opportunity to do research and providing invaluable guidance throughout this research. I would like to thank Brother Tomy Varghese as his dynamism, vision, and motivation have deeply inspired me.

Finally, my thanks go to all the people who have supported me to complete the research work directly or indirectly.

1

Conversion of Sound to Electric Energy

The "law of conservation of energy" states that energy can not be created nor be destroyed. below the thought of this law the technological giants have discovered varied supplys to extract energy from them and use it as a source of power for typical use. There square measure varied therefore referred to as eco-friendly sources of energy that we've discovered until the current artificial era. a number of them square measure enforced to nice extent below the acceptable circumstances to beat the short run of the energy thanks to technological boom that has light-emitting diode the energy must its apex. solar power is one within the list that came up with the big selection of applications like star heaters; star cookers and it gained success thanks to its simple implementation. There square measure varied alternative sources of renewable energy which incorporates harassing energy type wind, Biomass, water etc. however the potency of the energy sources mentioned on top of is that the major issue over that the scientists square measure operating since long. The potency of the cell is 2 hundredth solely below the sensible conditions. this is often not the sole downside with gift sources it any extends to high value concerned in production method. therefore the researchers currently square measure feeling the requirement of different kinds of sources to harass energy for our typical uses. to feature to the list there's Associate in Nursing rising state of affairs that leads America to a brand new renewable energy supply acknowledged to America since long which is that the sound. The sound or noise in alternative terms is gift all around America. therefore why not use it to satisfy our wants of energy. In our basic applications we have a tendency to see sound be regenerate within the electrical signals to travel over the media for communication functions. for instance the sound energy is regenerate into electrical signals victimization diaphragm gift within the electro-acoustic transducer and these signals then reach to the speakers so regenerate back to sound. The electrical current generated by a electro-acoustic transducer is extremely little and said as MIC-level; this signal is often measured in millivolts. Before it is used for all the world serious the signal must be amplified, sometimes to line level (typically zero.5 -2V).Application of sound energy because the supply of electricity is abundant helpful for the human existence as compared to alternative sources. this is often as a result of the sound is gift within the surroundings as a noise that forms a vital a part of the environmental pollution. The concentration of noise to use it for power generation will cause discovery of another hidden supply of energy which may act as a boon to non-renewable sources like coal, oil etc. that square measure on line of extinction.

Sound or popularly acknowledged to America as noise is one among the wide offered energy sources that have its vary extending virtually to eternity. The noise is taken into account to be a good contributor within the increasing pollution that is studied below the class of sound pollution. allow us to 1st perceive the essential definition of sound. Sound essentially is mechanical wave that's Associate in Nursing oscillation of pressure transmitted through some medium (like air or water), composed of frequencies that square measure among the vary of hearing. Thus, considering sound because the wave we are able to imagine it because the flow of energy from one purpose to a different with the assistance of a medium as air. The sound waves is longitudinal yet as thwartwise as per direction of vibration of the sound particles referred to as phonons. Sound that's perceptible by humans has frequencies from concerning twenty cps to twenty,000 Hz. In air at degree Celsius and pressure, the corresponding wavelengths of sound waves vary from seventeen m to seventeen millimetre. however have we have a tendency to ever imaginary sound as supply of electricity? No, is that the answer. this is often as a result of it absolutely was stone that was left right-side-out by the researchers up until currently however this hidden supply is currently rising because the a brand new era within the world of renewable sources of energy. this might be simply understood by the "law of thermodynamics" that states that the energy is regenerate to power.

In our day to day life we have a tendency to truly come upon varied devices that serve an equivalent purpose that's they convert the sound to electrical signals. [for example|for instance as Associate in Nursing example] a electrical device] is an example of a transducer, a tool that changes info from one type to a different. Sound info exists as patterns of air pressure; the electro-acoustic transducer changes this info into patterns of electrical current. The recording engineer is curious about the accuracy of this transformation, an idea he thinks of as fidelity. the essential plan is that sound is mechanical wave. once sound travels through any medium then it disturbs the particles of that specific medium and these disturbances caused by the sound is accustomed turn out electricity. The potency of the transducers and several other such devices is quiet low and can't be used for sensible applications. therefore the main arena to focus is however we are able to increase the potency of the electricity created by conversion of sound energy. allow us to currently see varied ways by that we are able to build a system to convert sound to power. the essential parameters that confirm the energy characteristics of noise square measure oscillation frequency and pressure level. Oscillation frequency is depicted in Hertz (Hz) and also the pressure level level is depicted by decibels (dB). Such electrical properties embrace Voltage (V), Current (I), resistance (R) and power (P). These quantities square measure associated with one another as, I=V/R, P=V^2/R.

In close to future if we have a tendency to square measure able to use this type of energy then it'll cause revolution within the field of the renewable sources of energy. thanks to development of recent sources like sound we are able to overcome the deficiency of electricity that we have a tendency to face within the developing countries across the globe. With the advancement of this technology we are able to conjointly imagine the charging of varied battery operated devices like our mobile phones simply by creating a decision to an admirer and talking. The mobile devices can virtually satisfy their name as they're going to become quiet moveable while not abundant concern concerning their battery life. Its alternative application field includes the lightening of the road lamps and traffic lights simply by extracting the sound energy of the noise that's created by the vehicles on the road. during this means we have a tendency to aren't solely able to cut back the sound pollution and however conjointly utilize it as a supply of electricity. conjointly within the industries with the mechanical forte wherever terribly vast quantity of the noise is created as results of functioning of serious machineries this sound is at bay and might be accustomed run the low power machines utilized in production method.

•   The sound energy is that the undiscovered supply that has huge potential to satisfy the longer term growing needs of the electricity and function the eco-friendly and renewable supply of energy.

•   This technology isn't much usable up until currently thanks to potency considerations however the current work on this field makes its future quiet promising.

•   Phonons square measure the particles of sound that offer the energy because the output that might be used for conversion as per the laws of physical science.

•   Present state of affairs states that researchers square measure endlessly attempting to evolve effective ways so as to boost its potency. On the premise of those works it is sure aforementioned that sound energy is that the successor of the renewable and ecofriendly sources of energy.

•   There square measure varied methodologies by that the sound is regenerate to electricity as

1.   methodology 1- This methodology is predicated on the faradays law of magnetic force induction and as per this methodology conversion of sound waves to electricity is done victimization diaphragm placed between magnetic poles.

2.   methodology 2- It illustrates the utilization novel technology that uses Piezo-electric materials to convert energy to electrical voltage. this sort of electricity is named as Piezo-electricity.

2

Turning Air pollution into Air Ink

———————◦ρ◦———————

"Pollution is nothing but the resources we are not harvesting. We allow them to disperse because we've been ignorant of their value. "
- Buckminster Fuller

Air Pollution includes the fine black particles principally composed of carbon, made by incomplete combustion of fossil fuels. The common attribute of soot particles is that they're extraordinarily small – two.5 micrometers or smaller in diameter. this can be smaller than dirt and mildew, and is concerning 1/30 the diameter of an individual's hair. It will travel deep into the respiratory organ, wherever the compounds it consists of will do some serious harm. UN agency report suggests that dirty air takes Brobdingnagian economic toll on poor countries and prices the globe quite $5 trillion annually in lost work days and welfare prices.

AIR-INKTM is associate ink whole that produces ink and ink-based art product by compression soot-based foamy effluents generated by automobiles because of incomplete combustion of fossil fuels. based by Graviky Labs, a product cluster of university Media research laboratory, Air Ink produces its materials through a piecemeal method that primarily involves capturing of emissions, separation of carbon from the soot, so compounding of this carbon with differing types of oils and solutions. Air Ink is marketed as an answer to pollution and its negative effects on human life.

Dubbed as "the initial ink created out of recycled pollution," its product were employed in August 2016 in association with Tiger brewage to form street art and murals in Hong Kong's Sheung Wan district.30–50 minutes of automobile pollution will provide enough carbon to fill one Air Ink pen. Anirudh Sharma, founding father of Graviky Labs, initial formed the concept of Air Ink in 2013 when he and his friends discovered that his garments were being stained by pollution. Sharma and his team spent on the brink of 3 years researching a way to purify and repurpose carbon soot from automotive vehicle emissions, a serious contributor to pollution

Soot composed of two.5-micrometer black carbon particles found in gasolene or diesel carbon emissions is captured from the tailpipes of cars through a tool referred to as 'Kaalink.' A separate method removes serious metals and transforms the carbon residue into paint or ink. one Air Ink pen contains 30–50 minutes of pollution. The emissions from two,500 hours of driving one commonplace diesel vehicle turn out concerning one hundred fifty litres of ink

Graviky Labs, the corporate behind Air Ink, claims the ink product will facilitate cut back transport pollution in cities. However, one in every of the company's lead engineers aforementioned, "Unless it's deployed on an outsized scale and there's a complete system around it, its impact are going to be borderline. Kaalink may be a cylindrical device that's retrofitted into a vehicle's system to gather the emissions. It will collect up to ninety three of the overall exhaust, that is then processed to get rid of serious metals and carcinogens.

The end-product from this device may be a refined carbon-based pigment. Kaalink has been tested on cars, trucks, motorcycles and fishing boats in urban center and metropolis. Kaalink is developed primarily for mitigating a number of the world's most harmful emissions. The device is ready well, and it's doing thus on atiny low scale. Kaalink is being employed to capture pollution from vehicle exhausts. it's then removing all deadly parts from the harvested pollution to form clean and safe black ink that artists will use as traditional ink.

According to the corporate "45 minutes value of transport emissions captured by the Kaalink device will turn out one fluid ounce of Air Ink." whereas cheaper carbon inks ar factory-made through the deliberate burning of fossil fuels, we have a tendency to use our proprietary device—what we have a tendency to decision KAALINK—to capture soot that's already being emitted from vehicles. KAALINK is retrofitted to the pipage of vehicles/generators to capture the outgoing pollutants. The unit mechanically activates once associate engine is activated and gases begin flowing through the exhaust. This activates the flow and thermo sensing element, which, in turn, engages a mechatronic capture system. All fine particle matter is then captured at intervals the walls of the unit. However, gases ar allowed to taste, feat the engine unaffected. once the lights on the outside of the unit flip from blue to red, the structure is full.

AIR-INK is presently on the market as 2mm, 15mm, 30mm and 50mm markers, and a 150ml screen ink set. With the success of campaign is AIR-INK can work towards emotional oil based mostly paints, cloth paints, outside paints, and more. AIR-INK follows industrial grade pointers, and artists are exploitation AIR-INK for the past half-dozen months while not problems. AIR-INK provides associate environmentally friendly choice for inventive and is top quality. The ink itself is thicker than most and might paint on rough surfaces while not haemorrhage. the color of the ink is solid black which inserts utterly with apply. On prime of that, AIR-INK presents an unbelievable chance to create an inventive community of personalities WHO hope to form the globe a bit a lot of property and delightful.

3

The Bermuda Triangle Truth

The Bermuda Triangle could be a stretch of the Atlantic Ocean deckle-edged by a line from Sunshine State to the islands of Bermuda, to Puerto anti-racketeering law, then back to Sunshine State. it's one among the largest mysteries of our time that won't extremely a mystery.

The term "Bermuda Triangle" was initial utilized in a writing written by Vincent H. Gaddis for fleet magazine in 1964. within the article, Gaddis claimed that during this strange ocean variety of ships and planes had disappeared while not clarification. Gaddis wasn't the primary one to return to the present conclusion, either. As early as 1952, George X. Sands -- in an exceedingly report in Fate magazine -- noted what looked like an oddly sizable amount of strange accidents in this region.

In 1969, John Wallace Spencer wrote a book known as Limbo of the Lost specifically concerning constellation. two years later, a feature documentary on the topic -- 'The Devil's Triangle' -- was discharged. These together with the bestseller The Bermuda Triangle (published in 1974) for good registered the legend of the "Hoodoo Sea" at intervals standard culture. many books instructed that the disappearances were because of AN intelligent, technologically advanced race living in house or underneath the ocean.

The Flight nineteen Mystery

The tale of 'Flight 19' started on December fifth, 1945. five attacker torpedo-bombers upraised into the air from the armed service base of operations at Fort Lauderdale, Sunshine State at 2:10 within the afternoon. it had been a routine observe mission. The flight was composed of all students apart from the commander (a Lt. Charles Taylor).

The mission needed Taylor and his cluster of thirteen men to fly cardinal compass point fifty six miles to Hens and Chicken Shoals to conduct observe bombing runs. once they had completed that objective, the flight arrange needed them to fly a further sixty seven miles East, then flip North for seventy three miles, and eventually straight back to base -- a distance of one hundred twenty miles. This course would take them on a triangular path over the ocean.

concerning AN hour-and-a-half once the flight had left, a Lt. Henry M. Robert Cox picked up a radio transmission from Taylor. Taylor indicated that his compasses weren't operating, however he believed himself to be somewhere over the coral reef (the 'Keys' ar an extended chain of islands south of the Sunshine State mainland). Cox urged him to fly north toward Miami; if Taylor was certain the flight was over the Keys.

Planes these days have variety of how that they will check their current position as well as being attentive to a collection of GPS (Global Positioning Satellites) in orbit round the Earth. it's nearly not possible for a pilot to urge lost if he has the proper instrumentation and uses it properly. In 1945, though, planes flying over water had to depend upon knowing their start line, however long and quick they'd flown, and in what direction. If a pilot created a blunder with any of those figures, he was lost. Over the ocean, there have been no landmarks to line him right.

Apparently Taylor had become confused at some purpose within the flight. He was AN seasoned pilot, however hadn't spent heaps of your time flying East toward the Bahamas that was wherever he was happening that day. for a few reason, Taylor apparently thought the flight had kicked off within the wrong direction and had headed south toward the Keys rather than East. This thought was to paint his selections throughout the remainder of the flight with deadly results.

The a lot of Taylor took his flight North to do to urge out of the Keys, the additional resolute ocean the Avengers really traveled. As time went on, snatches of transmissions were picked au courant the solid ground indicating the opposite Flight nineteen pilots were making an attempt to urge Taylor to alter course. "If we might simply fly west," one student told another, "we would get home." He was right.

By 4:45 PM, it had been obvious to the folks on the bottom that Taylor was dispiritedly lost. He was urged to show management of the flight over to at least one of his students, however apparently he did not. because it grew dark, communications deteriorated. From the few words that did get through, it had been apparent Taylor was still flying North and East -the wrong directions.

At 5:50 PM, the ComGulf ocean Frontier analysis Center managed get a fix on Flight 19's weakening signals. it had been apparently East of recent Izmir Beach, Florida. By then, communications were thus poor that this data couldn't be passed to the lost planes.

At 6:20, a "Dumbo Flying Boat" was sent to do and realize Flight nineteen and guide it back. at intervals the hour, two a lot of planes -- Martin Mariners -- joined the search. Hope was quickly weakening for Flight nineteen by then. The weather was obtaining rough and also the Avengers were terribly low on fuel.

The 2 Martin Mariners were presupposed to rendezvous at the search zone. The other -- selected 'Training 49' -- ne'er showed up. The last transmission from Flight nineteen was detected at 7:04 PM. Planes searched the world through the night and also the next day. There was no sign of the Avengers.

Nor did the authorities extremely expect to search out a lot of. The Avengers -- blinking once their fuel was exhausted -- would are sent to very cheap in seconds by the 50-foot waves of the storm. mutually of Taylor's colleagues noted, "they did not decision those planes 'Iron Birds' for nothing. They weighed fourteen,000 pounds empty. thus once they ditched, they went down pretty quick."

In 1991, five Avengers were found in 600 feet of water off the coast of Sunshine State by the salvage ship Deep ocean. Examination of the planes showed that they weren't Flight nineteen, however, therefore the final resting place of the planes and their crews remains the Bermuda Triangle's secret.

Many people believe that the Bermuda Triangle is placed more-or-less within the middle of a section of the Atlantic Ocean that when housed mythical place.

While several eminent Atlantean authorities specific differing opinions and ideas on exactly wherever mythical place was placed, i'd wish to sit down with a reading given by King of England Cayce in 1932:

"The position that the continent of mythical place occupied is between the Gulf of United Mexican States on the one hand and also the Mediterranean upon the opposite. Evidences of this lost civilization ar to be found within the mountain range and Morocco, Central American nation, Yucatan, and America. There ar some protrusive parts that has got to have at just the once or another been a little of this nice continent. British the Indies (or the Bahamas) ar a little of same that will be seen within the gift. If the geologic survey would be created in a number of these particularly -- or notably in Bimini and within the Gulf stream through this locality -- these could also be even nonetheless determined".

Mystery resolved

The Bermuda Triangle could be a mystery not. it's merely gas gas effervescent up from the once swamp (swamp gas) below the waters of that space. The void left behind swallows up airplanes and ships that ar ne'er found. $H2O$ is found that comes from the reaction of gas $CH4$ with salt water $NaCl$..

$CH4 + 4NaCl + 2H2$ zero $==> CCl4 + 4NaOH + O2$

The carbon within the gas takes up the chloride within the salt effort $H2O$ behind.

$CH4$ + metallic element $(OH) + 4HCl ==> CCl4 + 4NaOH ==> 4NaCl + 4OH$

$2CCl4 + 8NaOH ==> 8HCl + 2CO2 + 8NaCl + 2O2$

$8HCl + 8NaCl - 2CO2 + 2O2 ==> 2CCl4 + 8NaCl + 4O2 + 8H$

$4O2 + 24H2 ==> 4H2 O + 2O2$

(Fresh Water) There are four components water for each one-part gas. So, for one mole of gas, there's created four moles of $H2O$. Conclusion gas is a smaller amount dense than air. each ships and planes ar obsessed on fluid pressure to remain afloat. Drop the pressure underneath the wing of a device, the plane drops out of the air. drop the water pressure that floats a ship and also the boat sink. No remains ar found as a result of the gas gas leaves behind an oversized void. Mystery solved!

4

How Do Wind Turbines Work?

---◦▷◦---

Wind turbines work on an easy principle: rather than exploitation electricity to create wind—like a fan—wind turbines use wind to create electricity. Wind turns the propeller-like blades of a rotary engine around a rotor, that spins a generator, that creates electricity.

Wind could be a type of solar power caused by a mix of 3 cooccurring events:

1. The sun inconsistently heating the atmosphere
2. Irregularities of the surface
3. The rotation of the planet

Wind flow patterns and speeds vary greatly across the u. s. and area unit changed by bodies of water, vegetation, and variations in parcel of land. Humans use this wind flow, or motion energy, for several purposes: sailing, flying a kite, and even generating electricity.

The terms "wind energy" and "wind power" each describe the method by that the wind is employed to get mechanical power or electricity. This mechanical power will be used for specific tasks (such as grinding grain or pumping water) or a generator will convert this mechanical power into electricity.

A turbine turns wind energy into electricity exploitation the force from the rotor blades, that work like associate plane wing or whirlybird rotary wing. once wind flows across the blade, the gas pressure on one facet of the blade decreases. The distinction in gas pressure across the 2 sides of the blade creates each elevate and drag. The force of the elevate is stronger than the drag and this causes the rotor to spin. The rotor connects to the generator, either directly (if it's an instantaneous drive turbine) or through a shaft and a series of gears (a gearbox) that speed up the rotation and permit for a physically smaller generator. This translation of force to rotation of a generator creates electricity.

Types of Wind Turbines

Modern wind turbines be 2 basic groups:

HORIZONTAL-AXIS TURBINES

Horizontal-axis wind turbines area unit what many folks image once thinking of wind turbines. most typically, they need 3 blades and operate "upwind," with the rotary engine pivoting at the highest of the tower therefore the blades face into the wind.

VERTICAL-AXIS TURBINES

Vertical-axis wind turbines are available in many varieties, as well as the eggbeater-style Darrieus model, named when its French discoverer. These turbines area unit spatial relation, which means they don't have to be compelled to be adjusted to purpose into the wind to work.

Sizes of Wind Turbines

UTILITY-SCALE WIND TURBINES

Utility-scale wind turbines place size from one hundred kilowatts to as massive as many megawatts. Larger wind turbines area unit a lot of value effective and area unit sorted along into wind plants, which give bulk power to the electrical grid.

OFFSHORE WIND TURBINES

Offshore wind turbines tend to be huge, and taller than the sculpture of Liberty. they are doing not have constant transportation challenges of land-based wind installations, because the massive elements will be transported on ships rather than on roads. These turbines area unit ready to capture powerful ocean winds and generate large amounts of energy.

SINGLE little TURBINES

Single little turbines—below one hundred kilowatts—are usually used for residential, agricultural, and little industrial and industrial applications. little turbines will be employed in hybrid energy systems with alternative distributed energy resources, like small grids high-powered by diesel generators, batteries, and electrical phenomenon. These systems area unit known as hybrid wind systems and area unit usually employed in remote, off-grid locations(wherever a association to the utility grid isn't available) and have become a lot of common in grid-connected applications for resiliency.

DISTRIBUTED WIND

When wind turbines of any size area unit put in on the "customer" facet of the electrical meter, or area unit put in at or close to the place wherever the energy they manufacture are going to be used, they are known as "distributed wind.

Advantages and downsides of Wind-Generated Electricity

A Renewable Non-Polluting Resource

Wind energy could be a free, natural resource, thus notwithstanding what quantity is employed these days, there'll still be constant provide within the future. Wind energy is additionally a supply of unpolluted, non-polluting, electricity. not like standard power plants, wind plants emit no air pollutants or greenhouse gases. in line with the U.S. Department of Energy, in 1990, California's wind generation plants offset the emission of quite two.5 billion pounds of carbonic acid gas, and fifteen million pounds of alternative pollutants that will have otherwise been made. it'd take a forest of ninety million to a hundred seventy five million trees to produce constant air quality.

Cost problems

Even though the price of wind generation has weakened dramatically within the past ten years, the technology needs the next initial investment than fossil-fueled generators. Roughly eightieth of the price is that the machinery, with the balance being web site preparation and installation. If wind generating systems area unit compared with fossil-fueled systems on a "life-cycle" value basis (counting fuel and operational expenses for the lifetime of the generator), however, wind prices area unit rather more competitive with alternative generating technologies as a result of there's no fuel to buy and borderline operational expenses.

Environmental considerations

Although wind generation plants have comparatively very little impact on the atmosphere compared to fuel power plants, there's some concern over the noise made by the rotor blades, aesthetic (visual) impacts, and birds and kookie having been killed (avian/bat mortality) by flying into the rotors. Most of those issues are resolved or greatly reduced through technological development or by properly siting wind plants.

Supply and Transport problems

The major challenge to exploitation wind as a supply of power is that it's intermittent and doesn't continually blow once electricity is required. Wind can not be hold on (although wind-generated electricity will be hold on, if batteries area unit used), and not all winds will be controlled to fulfill the temporal order of electricity demands. Further, sensible wind sites area unit usually situated in remote locations off from areas of electrical power demand (such as cities). Finally, wind resource development might contend with alternative uses for the land, and people various uses is also a lot of extremely valued than electricity generation. However, wind turbines will be situated toward land that's additionally used for grazing or maybe farming.

5

Quantum Mechanics

—◆♌◆—

Quantum mechanics, science dealing with the behaviour of matter and light on the atomic and subatomic scale. It attempts to describe and account for the properties of molecules and atoms and their constituents—electrons, protons, neutrons, and other more esoteric particles such as quarks and gluons. These properties include the interactions of the particles with one another and with electromagnetic radiation (i.e., light, X-rays, and gamma rays).

The behaviour of matter and radiation on the atomic scale often seems peculiar, and the consequences of quantum theory are accordingly difficult to understand and to believe. Its concepts frequently conflict with common-sense notions derived from observations of the everyday world. There is no reason, however, why the behaviour of the atomic world should conform to that of the familiar, large-scale world. It is important to realize that quantum mechanics is a branch of physics and that the business of physics is to describe and account for the way the world—on both the large and the small scale—actually is and not how one imagines it or would like it to be.

The study of quantum mechanics is rewarding for several reasons. First, it illustrates the essential methodology of physics. Second, it has been enormously successful in giving correct results in practically every situation to which it has been applied. There is, however, an intriguing paradox. In spite of the overwhelming practical success of quantum mechanics, the foundations of the subject contain unresolved problems—in particular, problems concerning the nature of measurement. An essential feature of quantum mechanics is that it is generally impossible, even in principle, to measure a system without disturbing it; the detailed nature of this disturbance and the exact point at which it occurs are obscure and controversial. Thus, quantum mechanics attracted some of the ablest scientists of the 20th century, and they erected what is perhaps the finest intellectual edifice of the period.

Historical basis of quantum theory

Basic considerations

At a fundamental level, both radiation and matter have characteristics of particles and waves. The gradual recognition by scientists that radiation has particle-like properties and that matter has wavelike properties provided the impetus for the development of quantum mechanics. Influenced by Newton, most physicists of the 18th century believed that light consisted of particles, which they called corpuscles. From about 1800, evidence began to accumulate for a wave theory of light. At about this time Thomas Young showed that, if monochromatic light passes through a pair of slits, the two emerging beams interfere, so that a fringe pattern of alternately bright and dark bands appears on a screen. The bands are readily explained by a wave theory of light. According to the theory, a bright band is produced when the crests (and troughs) of the waves from the two slits arrive together at the screen; a dark band is produced when the crest of one wave arrives at the same time as the trough of the other, and the effects of the two light beams cancel. Beginning in 1815, a series of experiments by Augustin-Jean Fresnel of France and others showed that, when a parallel beam of light passes through a single slit, the emerging beam is no longer parallel but starts to diverge; this phenomenon is known as diffraction. Given the wavelength of the light and the geometry of the apparatus (i.e., the separation and widths of the slits and the distance from the slits to the screen), one can use the wave theory to calculate the expected pattern in each case; the theory agrees precisely with the experimental data.

Early developments

Planck's radiation law

By the end of the 19th century, physicists almost universally accepted the wave theory of light. However, though the ideas of classical physics explain interference and diffraction phenomena relating to the propagation of light, they do not account for the absorption and emission of light. All bodies radiate electromagnetic energy as heat; in fact, a body emits radiation at all wavelengths. The energy radiated at different wavelengths is a maximum at a wavelength that depends on the temperature of the body; the hotter the body, the shorter the wavelength for maximum radiation. Attempts to calculate the energy distribution for the radiation from a blackbody using classical ideas were unsuccessful. (A blackbody is a hypothetical ideal body or surface that absorbs and reemits all radiant energy falling on it.) One formula, proposed by Wilhelm Wien of Germany, did not agree with observations at long wavelengths, and another, proposed by Lord Rayleigh (John William Strutt) of England, disagreed with those at short wavelengths.

In 1900 the German theoretical physicist Max Planck made a bold suggestion. He assumed that the radiation energy is emitted, not continuously, but rather in discrete packets called quanta. The energy E of the quantum is related to the frequency v by $E = hv$. The quantity h, now known as Planck's constant, is a universal constant with the approximate value of 6.62607×10^{-34} joule·second. Planck showed that the calculated energy spectrum then agreed with observation over the entire wavelength range.

Einstein and the photoelectric effect

In 1905 Einstein extended Planck's hypothesis to explain the photoelectric effect, which is the emission of electrons by a metal surface when it is irradiated by light or more-energetic photons. The kinetic energy of the emitted electrons depends on the frequency v of the radiation, not on its intensity; for a given metal, there is a threshold frequency v_0 below which no electrons are emitted. Furthermore, emission takes place as soon as the light shines on the surface; there is no detectable delay. Einstein showed that these results can be explained by two assumptions: (1) that light is composed of corpuscles or photons, the energy of which is given by Planck's relationship, and (2) that an atom in the metal can absorb either a whole photon or nothing. Part of the energy of the absorbed photon frees an electron, which requires a fixed energy W, known as the work function of the metal; the rest is converted into the kinetic energy $m_e u^2/2$ of the emitted electron (m_e is the mass of the electron and u is its velocity). Thus, the energy relation is If v is less than v_0, where $hv_0 = W$, no electrons are emitted. Not all the experimental results mentioned above were known in 1905, but all Einstein's predictions have been verified since.

Bohr's theory of the atom

A major contribution to the subject was made by Niels Bohr of Denmark, who applied the quantum hypothesis to atomic spectra in 1913. The spectra of light emitted by gaseous atoms had been studied extensively since the mid-19th century. It was found that radiation from gaseous atoms at low pressure consists of a set of discrete wavelengths. This is quite unlike the radiation from a solid, which is distributed over a continuous range of wavelengths. The set of discrete wavelengths from gaseous atoms is known as a line spectrum, because the radiation (light) emitted consists of a series of sharp lines. The wavelengths of the lines are characteristic of the element and may form extremely complex patterns. The simplest spectra are those of atomic hydrogen and the alkali atoms (e.g., lithium, sodium, and potassium). For hydrogen the wavelengths λ are given by the empirical formula where m and n are positive integers with $n > m$ and R_∞, known as the Rydberg constant, has the value 1.097373157×10^7 per metre. For a given value of m, the lines for varying n form a series. The lines for $m = 1$, the Lyman series, lie in the ultraviolet part of the spectrum; those for $m = 2$, the Balmer series, lie in the visible spectrum; and those for $m = 3$, the Paschen series, lie in the infrared.

Bohr started with a model suggested by the New Zealand-born British physicist Ernest Rutherford. The model was based on the experiments of Hans Geiger and Ernest Marsden, who in 1909 bombarded gold atoms with massive, fast-moving alpha particles; when some of these particles were deflected backward, Rutherford concluded that the atom has a massive, charged nucleus. In Rutherford's model, the atom resembles a miniature solar system with the nucleus acting as the Sun and the electrons as the circulating planets. Bohr made three assumptions. First, he postulated that, in contrast to classical mechanics, where an infinite number of orbits is possible, an electron can be in only one of a discrete set of orbits, which he termed stationary states. Second, he postulated that the only orbits allowed are those for which the angular momentum of the electron is a whole number n times $\hbar$ ($\hbar = h/2\pi$). Third, Bohr assumed that Newton's laws of motion, so successful in calculating the paths of the planets around the Sun, also applied to electrons orbiting the nucleus. The force on the electron (the analogue of the gravitational force between the Sun and a planet) is the electrostatic attraction between the positively charged nucleus and the negatively charged electron. With these simple assumptions, he showed that the energy of the orbit has the form where E_0 is a constant that may be expressed by a combination of the known constants e, m_e, and $\hbar$. While in a stationary state the atom does not give off energy as light; however, when an electron makes a transition from a state with energy E_n to one with lower energy E_m, a quantum of energy is radiated with frequency ν, given by the equation Inserting the expression for E_n into this equation and using the relation $\lambda\nu = c$, where c is the speed of light, Bohr derived the formula for the wavelengths of the lines in the hydrogen spectrum, with the correct value of the Rydberg constant.

Bohr's theory was a brilliant step forward. Its two most important features have survived in present-day quantum mechanics. They are (1) the existence of stationary, nonradiating states and (2) the relationship of radiation frequency to the energy difference between the initial and final states in a transition. Prior to Bohr, physicists had thought that the radiation frequency would be the same as the electron's frequency of rotation in an orbit.

Scattering of X-rays

Soon scientists were faced with the fact that another form of radiation, X-rays, also exhibits both wave and particle properties. Max von Laue of Germany had shown in 1912 that crystals can be used as three-dimensional diffraction gratings for X-rays; his technique constituted the fundamental evidence for the wavelike nature of X-rays. The atoms of a crystal, which are arranged in a regular lattice, scatter the X-rays. For certain directions of scattering, all the crests of the X-rays coincide. (The scattered X-rays are said to be in phase and to give constructive interference.) For these directions, the scattered X-ray beam is very intense. Clearly, this phenomenon demonstrates wave behaviour. In fact, given the interatomic distances in the crystal and the directions of constructive interference, the wavelength of the waves can be calculated.

In 1922 the American physicist Arthur Holly Compton showed that X-rays scatter from electrons as if they are particles. Compton performed a series of experiments on the scattering of monochromatic, high-energy X-rays by graphite. He found that part of the scattered radiation had the same wavelength λ_0 as the incident X-rays but that there was an additional component with a longer wavelength λ. To interpret his results, Compton regarded the X-ray photon as a particle that collides and bounces off an electron in the graphite target as though the photon and the electron were a pair of (dissimilar) billiard balls. Application of the laws of conservation of energy and momentum to the collision leads to a specific relation between the amount of energy transferred to the electron and the angle of scattering. For X-rays scattered through an angle θ, the wavelengths λ and λ_0 are related by the equationThe experimental correctness of Compton's formula is direct evidence for the corpuscular behaviour of radiation.

De Broglie's wave hypothesis

Faced with evidence that electromagnetic radiation has both particle and wave characteristics, Louis-Victor de Broglie of France suggested a great unifying hypothesis in 1924. De Broglie proposed that matter has wave as well as particle properties. He suggested that material particles can behave as waves and that their wavelength λ is related to the linear momentum p of the particle by $\lambda = h/p$.

In 1927 Clinton Davisson and Lester Germer of the United States confirmed de Broglie's hypothesis for electrons. Using a crystal of nickel, they diffracted a beam of monoenergetic electrons and showed that the wavelength of the waves is related to the momentum of the electrons by the de Broglie equation. Since Davisson and Germer's investigation, similar experiments have been performed with atoms, molecules, neutrons, protons, and many other particles. All behave like waves with the same wavelength-momentum relationship.

6
Renewable Energy

Renewable energy, also called alternative energy, usable energy derived from replenishable sources such as the Sun (solar energy), wind (wind power), rivers (hydroelectric power), hot springs (geothermal energy), tides (tidal power), and biomass (biofuels).

At the beginning of the 21st century, about 80 percent of the world's energy supply was derived from fossil fuels such as coal, petroleum, and natural gas. Fossil fuels are finite resources; most estimates suggest that the proven reserves of oil are large enough to meet global demand at least until the middle of the 21st century. Fossil fuel combustion has a number of negative environmental consequences. Fossil-fueled power plants emit air pollutants such as sulfur dioxide, particulate matter, nitrogen oxides, and toxic chemicals (heavy metals: mercury, chromium, and arsenic), and mobile sources, such as fossil-fueled vehicles, emit nitrogen oxides, carbon monoxide, and particulate matter. Exposure to these pollutants can cause heart disease, asthma, and other human health problems. In addition, emissions from fossil fuel combustion are responsible for acid rain, which has led to the acidification of many lakes and consequent damage to aquatic life, leaf damage in many forests, and the production of smog in or near many urban areas. Furthermore, the burning of fossil fuels releases carbon dioxide (CO2), one of the main greenhouse gases that cause global warming.

In contrast, renewable energy sources accounted for nearly 20 percent of global energy consumption at the beginning of the 21st century, largely from traditional uses of biomass such as wood for heating and cooking. By 2015 about 16 percent of the world's total electricity came from large hydroelectric power plants, whereas other types of renewable energy (such as solar, wind, and geothermal) accounted for 6 percent of total electricity generation. Some energy analysts consider nuclear power to be a form of renewable energy because of its low carbon emissions; nuclear power generated 10.6 percent of the world's electricity in 2015.

Growth in wind power exceeded 20 percent and photovoltaics grew at 30 percent annually in the 1990s, and renewable energy technologies continued to expand throughout the early 21st century. Between 2001 and 2017 world total installed wind power capacity increased by a factor of 22, growing from 23,900 to 539,581 megawatts. Photovoltaic capacity also expanded, increasing by 50 percent in 2016 alone. The European Union (EU), which produced an estimated 6.38 percent of its energy from renewable sources in 2005, adopted a goal in 2007 to raise that figure to 20 percent by 2020. By 2016 some 17 percent of the EU's energy came from renewable sources. The goal also included plans to cut emissions of carbon dioxide by 20 percent and to ensure that 10 percent of all fuel consumption comes from biofuels. The EU was well on its way to achieving those targets by 2017. Between 1990 and 2016 the countries of the EU reduced carbon emissions by 23 percent and increased biofuel production to 5.5 percent of all fuels consumed in the region. In the United States numerous states have responded to concerns over climate change and reliance on imported fossil fuels by setting goals to increase renewable energy over time. For example, California required its major utility companies to produce 20 percent of their electricity from renewable sources by 2010, and by the end of that year California utilities were within 1 percent of the goal. In 2008 California increased this requirement to 33 percent by 2020, and in 2017 the state further increased its renewable-use target to 50 percent by 2030.

7
Dark Energy

---•♭•---

Dark energy, repulsive force that is the dominant component (69.4 percent) of the universe. The remaining portion of the universe consists of ordinary matter and dark matter. Dark energy, in contrast to both forms of matter, is relatively uniform in time and space and is gravitationally repulsive, not attractive, within the volume it occupies. The nature of dark energy is still not well understood.

A kind of cosmic repulsive force was first hypothesized by Albert Einstein in 1917 and was represented by a term, the "cosmological constant," that Einstein reluctantly introduced into his theory of general relativity in order to counteract the attractive force of gravity and account for a universe that was assumed to be static (neither expanding nor contracting). After the discovery in the 1920s by American astronomer Edwin Hubble that the universe is not static but is in fact expanding, Einstein referred to the addition of this constant as his "greatest blunder." However, the measured amount of matter in the mass-energy budget of the universe was improbably low, and thus some unknown "missing component," much like the cosmological constant, was required to make up the deficit. Direct evidence for the existence of this component, which was dubbed dark energy, was first presented in 1998.

Dark energy is detected by its effect on the rate at which the universe expands and its effect on the rate at which large-scale structures such as galaxies and clusters of galaxies form through gravitational instabilities. The measurement of the expansion rate requires the use of telescopes to measure the distance (or light travel time) of objects seen at different size scales (or redshifts) in the history of the universe. These efforts are generally limited by the difficulty in accurately measuring astronomical distances.

Since dark energy works against gravity, more dark energy accelerates the universe's expansion and retards the formation of large-scale structure. One technique for measuring the expansion rate is to observe the apparent brightness of objects of known luminosity like Type Ia supernovas. Dark energy was discovered in 1998 with this method by two international teams that included American astronomers Adam Riess (the author of this article) and Saul Perlmutter and Australian astronomer Brian Schmidt. The two teams used eight telescopes including those of the Keck Observatory and the MMT Observatory. Type Ia supernovas that exploded when the universe was only two-thirds of its present size were fainter and thus farther away than they would be in a universe without dark energy. This implied the expansion rate of the universe is faster now than it was in the past, a result of the current dominance of dark energy. (Dark energy was negligible in the early universe.)

Studying the effect of dark energy on large-scale structure involves measuring subtle distortions in the shapes of galaxies arising from the bending of space by intervening matter, a phenomenon known as "weak lensing." At some point in the last few billion years, dark energy became dominant in the universe and thus prevented more galaxies and clusters of galaxies from forming. This change in the structure of the universe is revealed by weak lensing. Another measure comes from counting the number of clusters of galaxies in the universe to measure the volume of space and the rate at which that volume is increasing. The goals of most observational studies of dark energy are to measure its equation of state (the ratio of its pressure to its energy density), variations in its properties, and the degree to which dark energy provides a complete description of gravitational physics.

In cosmological theory, dark energy is a general class of components in the stress-energy tensor of the field equations in Einstein's theory of general relativity. In this theory, there is a direct correspondence between the matter-energy of the universe (expressed in the tensor) and the shape of space-time. Both the matter (or energy) density (a positive quantity) and the internal pressure contribute to a component's gravitational field. While familiar components of the stress-energy tensor such as matter and radiation provide attractive gravity by bending space-time, dark energy causes repulsive gravity through negative internal pressure. If the ratio of the pressure to the energy density is less than $-1/3$, a possibility for a component with negative pressure, that component will be gravitationally self-repulsive. If such a component dominates the universe, it will accelerate the universe's expansion.

The simplest and oldest explanation for dark energy is that it is an energy density inherent to empty space, or a "vacuum energy." Mathematically, vacuum energy is equivalent to Einstein's cosmological constant. Despite the rejection of the cosmological constant by Einstein and others, the modern understanding of the vacuum, based on quantum field theory, is that vacuum energy arises naturally from the totality of quantum fluctuations (i.e., virtual particle-antiparticle pairs that come into existence and then annihilate each other shortly thereafter) in empty space. However, the observed density of the cosmological vacuum energy density is ~10–10 ergs per cubic centimetre; the value predicted from quantum field theory is ~10110 ergs per cubic centimetre. This discrepancy of 10120 was known even before the discovery of the far weaker dark energy. While a fundamental solution to this problem has not yet been found, probabilistic solutions have been posited, motivated by string theory and the possible existence of a large number of disconnected universes. In this paradigm the unexpectedly low value of the constant is understood as a result of an even greater number of opportunities (i.e., universes) for the occurrence of different values of the constant and the random selection of a value small enough to allow for the formation of galaxies (and thus stars and life).

Another popular theory for dark energy is that it is a transient vacuum energy resulting from the potential energy of a dynamical field. Known as "quintessence," this form of dark energy would vary in space and time, thus providing a possible way to distinguish it from a cosmological constant. It is also similar in mechanism (though vastly different in scale) to the scalar field energy invoked in the inflationary theory of the big bang.

Another possible explanation for dark energy is topological defects in the fabric of the universe. In the case of intrinsic defects in space-time (e.g., cosmic strings or walls), the production of new defects as the universe expands is mathematically similar to a cosmological constant, although the value of the equation of state for the defects depends on whether the defects are strings (one-dimensional) or walls (two-dimensional).

There have also been attempts to modify gravity to explain both cosmological and local observations without the need for dark energy. These attempts invoke departures from general relativity on scales of the entire observable universe.

A major challenge to understanding accelerated expansion with or without dark energy is to explain the relatively recent occurrence (in the past few billion years) of near-equality between the density of dark energy and dark matter even though they must have evolved differently. (For cosmic structures to have formed in the early universe, dark energy must have been an insignificant component.) This problem is known as the "coincidence problem" or the "fine-tuning problem." Understanding the nature of dark energy and its many related problems is one of the most formidable challenges in modern physics.

8
Electromagnetic Radiation

Electromagnetic radiation, in classical physics, the flow of energy at the universal speed of light through free space or through a material medium in the form of the electric and magnetic fields that make up electromagnetic waves such as radio waves, visible light, and gamma rays. In such a wave, time-varying electric and magnetic fields are mutually linked with each other at right angles and perpendicular to the direction of motion. An electromagnetic wave is characterized by its intensity and the frequency v of the time variation of the electric and magnetic fields.

In terms of the modern quantum theory, electromagnetic radiation is the flow of photons (also called light quanta) through space. Photons are packets of energy hv that always move with the universal speed of light. The symbol h is Planck's constant, while the value of v is the same as that of the frequency of the electromagnetic wave of classical theory. Photons having the same energy hv are all alike, and their number density corresponds to the intensity of the radiation. Electromagnetic radiation exhibits a multitude of phenomena as it interacts with charged particles in atoms, molecules, and larger objects of matter. These phenomena as well as the ways in which electromagnetic radiation is created and observed, the manner in which such radiation occurs in nature, and its technological uses depend on its frequency v. The spectrum of frequencies of electromagnetic radiation extends from very low values over the range of radio waves, television waves, and microwaves to visible light and beyond to the substantially higher values of ultraviolet light, X-rays, and gamma rays.

The basic properties and behaviour of electromagnetic radiation are discussed in this article, as are its various forms, including their sources, distinguishing characteristics, and practical applications. The article also traces the development of both the classical and quantum theories of radiation.

General considerations

Occurrence and importance

Close to 0.01 percent of the mass/energy of the entire universe occurs in the form of electromagnetic radiation. All human life is immersed in it, and modern communications technology and medical services are particularly dependent on one or another of its forms. In fact, all living things on Earth depend on the electromagnetic radiation received from the Sun and on the transformation of solar energy by photosynthesis into plant life or by biosynthesis into zooplankton, the basic step in the food chain in oceans. The eyes of many animals, including those of humans, are adapted to be sensitive to and hence to see the most abundant part of the Sun's electromagnetic radiation—namely, light, which comprises the visible portion of its wide range of frequencies. Green plants also have high sensitivity to the maximum intensity of solar electromagnetic radiation, which is absorbed by a substance called chlorophyll that is essential for plant growth via photosynthesis.

Practically all the fuels that modern society uses—gas, oil, and coal—are stored forms of energy received from the Sun as electromagnetic radiation millions of years ago. Only the energy from nuclear reactors does not originate from the Sun.

Everyday life is pervaded by artificially made electromagnetic radiation: food is heated in microwave ovens, airplanes are guided by radar waves, television sets receive electromagnetic waves transmitted by broadcasting stations, and infrared waves from heaters provide warmth. Infrared waves also are given off and received by automatic self-focusing cameras that electronically measure and set the correct distance to the object to be photographed. As soon as the Sun sets, incandescent or fluorescent lights are turned on to provide artificial illumination, and cities glow brightly with the colourful fluorescent and neon lamps of advertisement signs. Familiar too is ultraviolet radiation, which the eyes cannot see but whose effect is felt as pain from sunburn. Ultraviolet light represents a kind of electromagnetic radiation that can be harmful to life. Such is also true of X-rays, which are important in medicine as they allow physicians to observe the inner parts of the body but exposure to which should be kept to a minimum. Less familiar are gamma rays, which come from nuclear reactions and radioactive decay and are part of the harmful high-energy radiation of radioactive materials and nuclear weapons.

The electromagnetic spectrum

The brief account of familiar phenomena given above surveyed electromagnetic radiation from low frequencies of v (radio waves) to exceedingly high values of v (gamma rays). Going from the v values of radio waves to those of visible light is like comparing the thickness of this page with the distance of Earth from the Sun, which represents an increase by a factor of a million billion. Similarly, going from the v values of visible light to the very much larger ones of gamma rays represents another increase in frequency by a factor of a million billion. This extremely large range of v values, called the electromagnetic spectrum, is shown in Figure 1, together with the common names used for its various parts, or regions.

The number v is shared by both the classical and the modern interpretation of electromagnetic radiation. In classical language, v is the frequency of the temporal changes in an electromagnetic wave. The frequency of a wave is related to its speed c and wavelength λ in the following way. If 10 complete waves pass by in one second, one observes 10 wriggles, and one says that the frequency of such a wave is $v = 10$ cycles per second (10 hertz [Hz]). If the wavelength of the wave is, say, $\lambda = 3$ cm, then it is clear that a wave train 30 cm long has passed in that one second to produce the 10 wriggles that were observed. Thus, the speed of the wave is 30 cm per second, and one notes that in general the speed is $c = \lambda v$. The speed of electromagnetic radiation of all kinds is the same universal constant that is defined to be exactly $c = 299{,}792{,}458$ metres per second (186,282 miles per second). The wavelengths of the classical electromagnetic waves in free space calculated from $c = \lambda v$ are also shown on the spectrum in Figure 1, as is the energy hv of modern-day photons. Commonly used as the unit of energy is the electron volt (eV), which is the energy that can be given to an electron by a one-volt battery. It is clear that the range of wavelengths λ and of photon energies hv are equally as large as the spectrum of v values.

Because the wavelengths and energy quanta hv of electromagnetic radiation of the various parts of the spectrum are so different in magnitude, the sources of the radiations, the interactions with matter, and the detectors employed are correspondingly different. This is why the same electromagnetic radiation is called by different names in various regions of the spectrum.

In spite of these obvious differences of scale, all forms of electromagnetic radiation obey certain general rules that are well understood and that allow one to calculate with very high precision their properties and interactions with charged particles in atoms, molecules, and large objects. Electromagnetic radiation is, classically speaking, a wave of electric and magnetic fields propagating at the speed of light c through empty space. In this wave the electric and magnetic fields change their magnitude and direction each second. This rate of change is the frequency v measured in cycles per second—namely, in hertz. The electric and magnetic fields are always perpendicular to each other and at right angles to the direction of propagation, as shown in Figure 2. There is as much energy carried by the electric component of the wave as by the magnetic component, and the energy is proportional to the square of the field strength.

Generation of electromagnetic radiation

Electromagnetic radiation is produced whenever a charged particle, such as an electron, changes its velocity—i.e., whenever it is accelerated or decelerated. The energy of the electromagnetic radiation thus produced comes from the charged particle and is therefore lost by it. A common example of this phenomenon is the oscillating charge or current in a radio antenna. The antenna of a radio transmitter is part of an electric resonance circuit in which the charge is made to oscillate at a desired frequency. An electromagnetic wave so generated can be received by a similar antenna connected to an oscillating electric circuit in the tuner that is tuned to that same frequency. The electromagnetic wave in turn produces an oscillating motion of charge in the receiving antenna. In general, one can say that any system which emits electromagnetic radiation of a given frequency can absorb radiation of the same frequency.

Such human-made transmitters and receivers become smaller with decreasing wavelength of the electromagnetic wave and prove impractical in the millimetre range. At even shorter wavelengths down to the wavelengths of X-rays, which are one million times smaller, the oscillating charges arise from moving charges in molecules and atoms.

One may classify the generation of electromagnetic radiation into two categories: (1) systems or processes that produce radiation covering a broad continuous spectrum of frequencies and (2) those that emit (and absorb) radiation of discrete frequencies that are characteristic of particular systems. The Sun with its continuous spectrum is an example of the first, while a radio transmitter tuned to one frequency exemplifies the second category.

Continuous spectra of electromagnetic radiation

Such spectra are emitted by any warm substance. Heat is the irregular motion of electrons, atoms, and molecules; the higher the temperature, the more rapid the motion. Since electrons are much lighter than atoms, irregular thermal motion produces irregular oscillatory charge motion, which reflects a continuous spectrum of frequencies. Each oscillation at a particular frequency can be considered a tiny "antenna" that emits and receives electromagnetic radiation. As a piece of iron is heated to increasingly high temperatures, it first glows red, then yellow, and finally white. In short, all the colours of the visible spectrum are represented. Even before the iron begins to glow red, one can feel the emission of infrared waves by the heat sensation on the skin. A white-hot piece of iron also emits ultraviolet radiation, which can be detected by a photographic film.

Not all materials heated to the same temperature emit the same amount and spectral distribution of electromagnetic waves. For example, a piece of glass heated next to iron looks nearly colourless, but it feels hotter to the skin (it emits more infrared rays) than does the iron. This observation illustrates the rule of reciprocity: a body radiates strongly at those frequencies that it is able to absorb, because for both processes it needs the tiny antennas of that range of frequencies. Glass is transparent in the visible range of light because it lacks possible electronic absorption at these particular frequencies. As a consequence, glass cannot glow red because it cannot absorb red. On the other hand, glass is a better emitter/absorber in the infrared than iron or any other metal that strongly reflects such lower-frequency electromagnetic waves. This selective emissivity and absorptivity is important for understanding the greenhouse effect (see below The greenhouse effect of the atmosphere) and many other phenomena in nature. The tungsten filament of a lightbulb has a temperature of 2,500 K (4,040 °F) and emits large amounts of visible light but relatively little infrared because metals, as mentioned above, have small emissivities in the infrared range. This is of course fortunate, since one wants light from a lightbulb but not much heat. The light emitted by a candle originates from very hot carbon soot particles in the flame, which strongly absorb and thus emit visible light. By contrast, the gas flame of a kitchen range is pale, even though it is hotter than a candle flame, because of the absence of soot. Light from the stars originates from the high temperature of the gases at their surface. A wide spectrum of radiation is emitted from the Sun's surface, the temperature of which is about 5,800 K. The radiation output is 60 million watts for every square metre of solar surface, which is equivalent to the amount produced by an average-size commercial power-generating station that can supply electric power for about 30,000 households.

The spectral composition of a heated body depends on the materials of which the body consists. That is not the case for an ideal radiator or absorber. Such an ideal object absorbs and thus emits radiation of all frequencies equally and fully. A radiator/absorber of this kind is called a blackbody, and its radiation spectrum is referred to as blackbody radiation, which depends on only one parameter, its temperature. Scientists devise and study such ideal objects because their properties can be known exactly. This information can then be used to determine and understand why real objects, such as a piece of iron or glass, a cloud, or a star, behave differently.

A good approximation of a blackbody is a piece of coal or, better yet, a cavity in a piece of coal that is visible through a small opening. There is one property of blackbody radiation which is familiar to everyone but which is actually quite mysterious. As the piece of coal is heated to higher and higher temperatures, one first observes a dull red glow, followed by a change in colour to bright red; as the temperature is increased further, the colour changes to yellow and finally to white. White is not itself a colour but rather the visual effect of the combination of all primary colours. The fact that white glow is observed at high temperatures means that the colour blue has been added to the ones observed at lower temperatures. This colour change with temperature is mysterious because one would expect, as the energy (or temperature) is increased, just more of the same and not something entirely different. For example, as one increases the power of a radio amplifier, one hears the music louder but not at a higher pitch.

The change in colour or frequency distribution of the electromagnetic radiation coming from heated bodies at different temperatures remained an enigma for centuries. The solution of this mystery by the German physicist Max Planck initiated the era of modern physics at the beginning of the 20th century. He explained the phenomenon by proposing that the tiny antennas in the heated body are quantized, meaning that they can emit electromagnetic radiation only in finite energy quanta of size hv. The universal constant h is called Planck's constant in his honour. For blue light hv = 3 eV, whereas hv = 1.8 eV for red light. Since high-frequency antennas of vibrating charges in solids have to emit larger energy quanta hv than lower-frequency antennas, they can only do so when the temperature, or the thermal atomic motion, becomes high enough. Hence, the average pitch, or peak frequency, of blackbody electromagnetic radiation increases with temperature.

The many tiny antennas in a heated chunk of material are, as noted above, to be identified with the accelerating and decelerating charges in the heat motion of the atoms of the material. There are other sources of continuous spectra of electromagnetic radiation that are not associated with heat but still come from accelerated or decelerated charges. X-rays are, for example, produced by abruptly stopping rapidly moving electrons. This deceleration of the charges produces bremsstrahlung ("braking radiation"). In an X-ray tube, electrons moving with an energy of E_{max} = 10,000 to 50,000 eV (10–50 keV) are made to strike a piece of metal. The electromagnetic radiation produced by this sudden deceleration of electrons is a continuous spectrum extending up to the maximum photon energy $h\nu = E_{max}$.

By far the brightest continuum spectra of electromagnetic radiation come from synchrotron radiation sources. These are not well known because they are predominantly used for research and sometimes for commercial and medical applications. Because any change in motion is an acceleration, circulating currents of electrons produce electromagnetic radiation. When these circulating electrons move at relativistic speeds (i.e., those approaching the speed of light), the brightness of the radiation increases enormously. This radiation was first observed at the General Electric Company in 1947 in an electron synchrotron (hence the name of this radiation), which is a type of particle accelerator that forces relativistic electrons into circular orbits by using powerful magnetic fields. The intensity of synchrotron radiation is further increased more than a thousandfold by wigglers and undulators that move the beam of relativistic electrons to and fro by means of other magnetic fields.

The conditions for generating bremsstrahlung as well as synchrotron radiation exist in nature in various forms. Acceleration and capture of charged particles by the gravitational field of a star, black hole, or galaxy is a source of energetic cosmic X-rays. Gamma rays are produced in other kinds of cosmic objects—namely, supernovae, neutron stars, and quasars.

Discrete-frequency sources and absorbers of electromagnetic radiation

These are commonly encountered in everyday life. Familiar examples of discrete-frequency electromagnetic radiation include the distinct colours of lamps filled with different fluorescent gases that are characteristic of advertisement signs, the colours of dyes and pigments, the bright yellow of sodium lamps, the blue-green hue of mercury lamps, and the specific colours of lasers.

Sources of electromagnetic radiation of specific frequency are typically atoms or molecules. Every atom or molecule can have certain discrete internal energies, which are called quantum states. An atom or molecule can therefore change its internal energy only by discrete amounts. By going from a higher to a lower energy state, a quantum hv of electromagnetic radiation is emitted of a magnitude that is precisely the energy difference between the higher and lower state. Absorption of a quantum hv takes the atom from a lower to a higher state if hv matches the energy difference. All like atoms are identical, but each chemical element of the periodic table has its own specific set of possible internal energies. Therefore, by measuring the characteristic and discrete electromagnetic radiation that is either emitted or absorbed by atoms or molecules, one can identify which kind of atom or molecule is giving off or absorbing the radiation. This provides a means of determining the chemical composition of substances. Since one cannot subject a piece of a distant star to conventional chemical analysis, studying the emission or absorption of starlight is the only way to determine the composition of stars or of interstellar gases and dust..

The Sun, for example, not only emits the continuous spectrum of radiation that originates from its hot surface but also emits discrete radiation quanta hv that are characteristic of its atomic composition. Many of the elements can be detected at the solar surface, but the most abundant is helium. This is so because helium is the end product of the nuclear fusion reaction that is the fundamental energy source of the Sun. This particular element was named helium (from the Greek word helios, meaning "Sun") because its existence was first discovered by its characteristic absorption energies in the Sun's spectrum. The helium of the cooler outer parts of the solar atmosphere absorbs the characteristic light frequencies from the lower and hotter regions of the Sun.

The characteristic and discrete energies hv found as emission and absorption of electromagnetic radiation by atoms and molecules extend to X-ray energies. As high-energy electrons strike the piece of metal in an X-ray tube, electrons are knocked out of the inner energy shell of the atoms. These vacancies are then filled by electrons from the second or third shell; emitted in the process are X-rays having hv values that correspond to the energy differences of the shells. One therefore observes not only the continuous spectrum of the bremsstrahlung discussed above but also X-ray emissions of discrete energies hv that are characteristic of the specific elemental composition of the metal struck by the energetic electrons in the X-ray tube.

The discrete electromagnetic radiation energies hv emitted or absorbed by all substances reflect the discreteness of the internal energies of all material things. This means that window glass and water are transparent to visible light; they cannot absorb these visible light quanta because their internal energies are such that no energy difference between a higher and a lower internal state matches the energy hv of visible light. Figure 3 shows as an example the coefficient of absorption of water as a function of frequency v of electromagnetic radiation. Above the scale of frequencies, the corresponding scales of photon energy hv and wavelength λ are given. An absorption coefficient α = 10–4 cm–1 means that the intensity of electromagnetic radiation is only one-third its original value after passing through 100 metres of water. When α = 1 cm–1, only a layer 1 cm thick is needed to decrease the intensity to one-third its original value, and, for α = 103 cm, a layer of water having the thickness of a thin sheet of paper is sufficient to attenuate electromagnetic radiation by that much. The transparency of water to visible light, marked by the vertical dashed lines, is a remarkable feature that is significant for life on Earth.

All things look so different and have different colours because of their different sets of internal discrete energies, which determine their interaction with electromagnetic radiation. The words looking and colours are associated with the human detectors of electromagnetic radiation, the eyes. Since there are instruments available for detecting electromagnetic radiation of any frequency, one can imagine that things "look" different at different energies of the spectrum because different materials have their own characteristic sets of discrete internal energies. Even the nuclei of atoms are composites of other elementary particles and thus can be excited to many discrete internal energy states. Since nuclear energies are much larger than atomic energies, the energy differences between internal energy states are substantially larger, and the corresponding electromagnetic radiation quanta hv emitted or absorbed when nuclei change their energies are even bigger than those of X-rays. Such quanta given off or absorbed by atomic nuclei are called gamma rays (see above The electromagnetic spectrum).

Properties and behaviour

Scattering, reflection, and refraction

If a charged particle interacts with an electromagnetic wave, it experiences a force proportional to the strength of the electric field and thus is forced to change its motion in accordance with the frequency of the electric field wave. In doing so, it becomes a source of electromagnetic radiation of the same frequency, as described in the previous section. The energy for the work done in accelerating the charged particle and emitting this secondary radiation comes from and is lost by the primary wave. This process is called scattering.

Since the energy density of the electromagnetic radiation is proportional to the square of the electric field strength and the field strength is caused by acceleration of a charge, the energy radiated by such a charge oscillator increases with the square of the acceleration. On the other hand, the acceleration of an oscillator depends on the frequency of the back-and-forth oscillation. The acceleration increases with the square of the frequency. This leads to the important result that the electromagnetic energy radiated by an oscillator increases very rapidly—namely, with the square of the square or, as one says, with the fourth power of the frequency. Doubling the frequency thus produces an increase in radiated energy by a factor of 16.

This rapid increase in scattering with the frequency of electromagnetic radiation can be seen on any sunny day: it is the reason the sky is blue and the setting Sun is red. The higher-frequency blue light from the Sun is scattered much more by the atoms and molecules of Earth's atmosphere than is the lower-frequency red light. Hence, the light of the setting Sun, which passes through a thick layer of atmosphere, has much more red than yellow or blue light, while light scattered from the sky contains much more blue than yellow or red light.

The process of scattering, or reradiating part of the electromagnetic wave by a charge oscillator, is fundamental to understanding the interaction of electromagnetic radiation with solids, liquids, or any matter that contains a very large number of charges and thus an enormous number of charge oscillators. This also explains why a substance that has charge oscillators of certain frequencies absorbs and emits radiation of those frequencies.

When electromagnetic radiation falls on a large collection of individual small charge oscillators, as in a piece of glass or metal or a brick wall, all of these oscillators perform oscillations in unison, following the beat of the electric wave. As a result, all the oscillators emit secondary radiation in unison (or coherently), and the total secondary radiation coming from the solid consists of the sum of all these secondary coherent electromagnetic waves. This sum total yields radiation that is reflected from the surface of the solid and radiation that goes into the solid at a certain angle with respect to the normal of (i.e., a line perpendicular to) the surface. The latter is the refracted radiation that may be attenuated (absorbed) on its way through the solid.

Superposition and interference

When two electromagnetic waves of the same frequency superpose in space, the resultant electric and magnetic field strength of any point of space and time is the sum of the respective fields of the two waves. When one forms the sum, both the magnitude and the direction of the fields need be considered, which means that they sum like vectors. In the special case when two equally strong waves have their fields in the same direction in space and time (i.e., when they are in phase), the resultant field is twice that of each individual wave. The resultant intensity, being proportional to the square of the field strength, is therefore not two but four times the intensity of each of the two superposing waves.

By contrast, the superposition of a wave that has an electric field in one direction (positive) in space and time with a wave of the same frequency having an electric field in the opposite direction (negative) in space and time leads to cancellation and no resultant wave at all (zero intensity). Two waves of this sort are termed out of phase. The first example, that of in-phase superposition yielding four times the individual intensity, constitutes what is called constructive interference. The second example, that of out-of-phase superposition yielding zero intensity, is destructive interference. Since the resultant field at any point and time is the sum of all individual fields at that point and time, these arguments are easily extended to any number of superposing waves. One finds constructive, destructive, or partial interference for waves having the same frequency and given phase relationships.

Propagation and coherence

Once generated, an electromagnetic wave is self-propagating because a time-varying electric field produces a time-varying magnetic field and vice versa. When an oscillating current in an antenna is switched on for, say, eight minutes, then the beginning of the electromagnetic train reaches the Sun just when the antenna is switched off because it takes a few seconds more than eight minutes for electromagnetic radiation to reach the Sun. This eight-minute wave train, which is as long as the Sun–Earth distance, then continues to travel with the speed of light past the Sun into the space beyond.

Except for radio waves transmitted by antennas that are switched on for many hours, most electromagnetic waves comes in many small pieces. The length and duration of a wave train are called coherence length and coherence time, respectively. Light from the Sun or from a lightbulb comes in many tiny bursts lasting about a millionth of a millionth of a second and having a coherence length of about one centimetre. The discrete radiant energy emitted by an atom as it changes its internal energy can have a coherence length several hundred times longer (one to 10 metres) unless the radiating atom is disturbed by a collision.

The time and space at which the electric and magnetic fields have a maximum value or are zero between the reversal of their directions are different for different wave trains. It is therefore clear that the phenomenon of interference can arise only from the superposition of part of a wave train with itself. This can be accomplished, for instance, with a half-transparent mirror that reflects half the intensity and transmits the other half of each of the billion billion wave trains of a given light source, say, a yellow sodium discharge lamp. One can allow one of these half beams to travel in direction A and the other in direction B, as shown in Figure 4. By reflecting each half beam back, one can then superpose the two half beams and observe the resultant total. If one half beam has to travel a path 1/2 wavelength or 3/2 or 5/2 wavelength longer than the other, then the superposition yields no light at all because the electric and magnetic fields of every half wave train in the two half beams point in opposite directions and their sum is therefore zero. The important point is that cancellation occurs between each half wave train and its mate. This is an example of destructive interference. By adjusting the path lengths A and B such that they are equal or differ by λ, 2λ, 3λ..., the electric and magnetic fields of each half wave train and its mate add when they are superposed. This is constructive interference, and, as a result, one sees strong light.

Speed of electromagnetic radiation and the Doppler effect

Electromagnetic radiation—or, in modern terminology, the photons hν—always travels in free space with the universal speed c—i.e., the speed of light. This is actually a very puzzling situation which was first experimentally verified by Michelson and Edward Williams Morley, another American scientist, in 1887. It is the basic axiom of Albert Einstein's theory of relativity. Although there is no doubt that it is true, the situation is puzzling because it is so different from the behaviour of normal particles—that is to say, for little or not so little pieces of matter. When one chases behind a normal particle (e.g., an airplane) or moves from the opposite direction toward it, one certainly will measure very different speeds of the airplane relative to oneself. One would detect a very low relative speed in the first case and a very high one in the second. Moreover, a bullet shot forward from the airplane and another toward the back would appear to be moving with different speeds relative to oneself. This is not at all the case when one measures the speed of electromagnetic radiation: irrespective of one's motion or that of the source of the electromagnetic radiation, any measurement by a moving observer will result in the universal speed of light. This must be accepted as a fact of nature.

What happens to pitch or frequency when the source is moving toward the observers or away from them? It has been established from sound waves that the frequency is higher when a sound source is moving toward the observers and lower when it is moving away from them. This is the Doppler effect, named after the Austrian physicist Christian Doppler, who first described the phenomenon in 1842. Doppler predicted that the effect also occurs with electromagnetic radiation and suggested that it be used for measuring the relative speeds of stars. This explains why a characteristic blue light emitted, for example, by an excited helium atom as it changes from a higher to a lower internal energy state no longer appears blue when one looks at this light coming from helium atoms that move very rapidly away from Earth with, say, a galaxy. When the speed of such a galaxy away from Earth is high, the light may appear yellow; if the speed is still higher, it may appear red or even infrared. Hence, the speed of galaxies as well as of stars relative to Earth is measured from the Doppler shift of characteristic atomic radiation energies $h\nu$.

Cosmic background electromagnetic radiation

As one measures the relative speeds of galaxies by using the Doppler shift of characteristic radiation emissions, one finds that all galaxies are moving away from one another. Those that are moving the fastest are systems that are the farthest away (Hubble's law). The speeds and distances give the appearance of an explosion. This explosion, dubbed the big bang, is calculated to have occurred 13.8 billion years ago, which is considered to be the age of the universe. From this early stage onward, the universe expanded and cooled. The American scientists Robert W. Wilson and Arno Penzias determined in 1965 that the whole universe can be conceived of as an expanding blackbody filled with electromagnetic radiation which now corresponds to a temperature of 2.735 K, only a few degrees above absolute zero. Because of this low temperature, most of the radiation energy is in the microwave region of the electromagnetic spectrum. The intensity of this radiation corresponds, on average, to about 400 photons in every cubic centimetre of the universe. It has been estimated that there are about one billion times more photons in the universe than electrons, nuclei, and all other things taken together. The presence of this microwave cosmic background radiation supports the predictions of big-bang cosmology.

Wilkinson Microwave Anisotropy ProbeA full-sky map produced by the Wilkinson Microwave Anisotropy Probe (WMAP) showing cosmic background radiation, a very uniform glow of microwaves emitted by the infant universe more than 13 billion years ago. Colour differences indicate tiny fluctuations in the intensity of the radiation, a result of tiny variations in the density of matter in the early universe. According to inflation theory, these irregularities were the "seeds" that became the galaxies. WMAP's data support the big bang and inflation models.NASA/WMAP Science Team

Effect of gravitation

The energy of the quanta of electromagnetic radiation is subject to gravitational forces just like a mass of magnitude $m = hv/c^2$. This is so because the relationship of energy E and mass m is $E = mc^2$. As a consequence, light traveling toward Earth gains energy and its frequency is shifted toward the blue (shorter wavelengths), whereas light traveling "up" loses energy and its frequency is shifted toward the red (longer wavelengths). These shifts are very small but have been detected by the American physicists Robert V. Pound and Glen A. Rebka.

The effect of gravitation on light increases with the strength of the gravitational attraction. Thus, a light beam from a distant star does not travel along a straight line when passing a star like the Sun but is deflected toward it. This deflection can be strong around very heavy cosmic objects, which then distort the light path acting as a gravitational lens.

Under extreme conditions the gravitational force of a cosmic object can be so strong that no electromagnetic radiation can escape the gravitational pull. Such an object, called a black hole, is therefore not visible, and its presence can be detected only by its gravitational effect on other, visible objects in its vicinity. (For additional information, see astronomy.)

The greenhouse effect of the atmosphere

The temperature of the terrestrial surface environment is controlled not only by the Sun's electromagnetic radiation but also in a sensitive way by

Earth's atmosphere. As noted earlier, each substance absorbs and emits electromagnetic radiation of some energies hv and does not do so in other ranges of energy. These regions of transparency and opaqueness are governed by the particular distribution of internal energies of the substance.

Earth's atmosphere acts much like the glass panes of a greenhouse: it allows sunlight, particularly its visible range, to reach and warm Earth, but it largely inhibits the infrared radiation emitted by the heated terrestrial surface from escaping into space. Since the atmosphere becomes thinner and thinner with increasing altitude above Earth, there is less atmospheric absorption in the higher regions of the atmosphere. At an altitude of 100 km (62 miles), the fraction of atmosphere is one 10-millionth of that on the ground. Below 10 million hertz (107 Hz), the absorption is caused by the ionosphere, a layer in which atoms and molecules in the atmosphere are ionized by the Sun's ultraviolet radiation. In the infrared region, the absorption is caused by molecular vibrations and rotations. In the ultraviolet and X-ray regions, the absorption is due to electronic excitations in atoms and molecules.

greenhouse effect on EarthThe greenhouse effect on Earth. Some incoming sunlight is reflected by Earth's atmosphere and surface, but most is absorbed by the surface, which is warmed. Infrared (IR) radiation is then emitted from the surface. Some IR radiation escapes to space, but some is absorbed by the atmosphere's greenhouse gases (especially water vapour, carbon dioxide, and methane) and reradiated in all directions, some to space and some back toward the surface, where it further warms the surface and the lower atmosphere.Encyclopædia Britannica, Inc.

Without water vapour and carbon dioxide (CO_2), which are, together with certain industrial pollutants, the main infrared-absorbing species in the atmosphere, Earth would experience the extreme temperature variations between night and day that occur on the Moon. Earth would then be a frozen planet, like Mars, with an average temperature of 200 K (–73 °C, or –100 °F), and not be able to support life. Scientists believe that Earth's temperature and climate in general will be affected as the composition of the atmosphere is altered by an increased release and accumulation of carbon dioxide and other gaseous pollutants (for a detailed discussion, see climate; hydrosphere; and global warming).

Forms of electromagnetic radiation

Electromagnetic radiation appears in a wide variety of forms and manifestations. Yet, these diverse phenomena are understood to comprise a single aspect of nature, following simple physical principles. Common to all forms is the fact that electromagnetic radiation interacts with and is generated by electric charges. The apparent differences in the phenomena arise from the question in which environment and under what circumstances can charges respond on the time scale of the frequency v of the radiation.

At smaller frequencies v (smaller than 1012 hertz), electric charges typically are the freely moving electrons in the metal components of antennas or the free electrons and ions in space that give rise to phenomena related to radio waves, radar waves, and microwaves. At higher frequencies (1012 to 5 × 1014 hertz), in the infrared region of the spectrum, the moving charges are primarily associated with the rotations and vibrations of molecules and the motions of atoms bonded together in materials. Electromagnetic radiation in the visible range to X-rays have frequencies that correspond to charges within atoms, whereas gamma rays are associated with frequencies of charges within atomic nuclei. The characteristics of electromagnetic radiation occurring in the different regions of the spectrum are described in this section.

Radio waves

Radio waves are used for wireless transmission of sound messages, or information, for communication, as well as for maritime and aircraft navigation. The information is imposed on the electromagnetic carrier wave as amplitude modulation (AM) or as frequency modulation (FM) or in digital form (pulse modulation). Transmission therefore involves not a single-frequency electromagnetic wave but rather a frequency band whose width is proportional to the information density. The width is about 10,000 Hz for telephone, 20,000 Hz for high-fidelity sound, and five megahertz (MHz = one million hertz) for high-definition television. This width and the decrease in efficiency of generating electromagnetic waves with decreasing frequency sets a lower frequency limit for radio waves near 10,000 Hz.

Because electromagnetic radiation travels in free space in straight lines, late 19th-century scientists questioned the efforts of the Italian physicist and inventor Guglielmo Marconi to develop long-range radio. Earth's curvature limits the line-of-sight distance from the top of a 100-metre (330-foot) tower to about 30 km (19 miles). Marconi's unexpected success in transmitting messages over more than 2,000 km (1,200 miles) led to the discovery of the Kennelly-Heaviside layer, more commonly known as the ionosphere. This region is an approximately 300-km- (190-mile-) thick layer starting about

100 km (60 miles) above Earth's surface in which the atmosphere is partially ionized by ultraviolet light from the Sun, giving rise to enough electrons and ions to affect radio waves. Because of the Sun's involvement, the height, width, and degree of ionization of the stratified ionosphere vary from day to night and from summer to winter.

Radio waves transmitted by antennas in certain directions are bent or even reflected back to Earth by the ionosphere, as illustrated in Figure 5. They may bounce off Earth and be reflected by the ionosphere repeatedly, making radio transmission around the globe possible. Long-distance communication is further facilitated by the so-called ground wave. This form of electromagnetic wave closely follows Earth's surface, particularly over water, as a result of the wave's interaction with the terrestrial surface. The range of the ground wave (up to 1,600 km [1,000 miles]) and the bending and reflection of the sky wave by the ionosphere depend on the frequency of the waves. Under normal ionospheric conditions 40 MHz is the highest-frequency radio wave that can be reflected from the ionosphere. In order to accommodate the large band width of transmitted signals, television frequencies are necessarily higher than 40 MHz. Television transmitters must therefore be placed on high towers or on hilltops.

As a radio wave travels from the transmitting to the receiving antenna, it may be disturbed by reflections from buildings and other large obstacles. Disturbances arise when several such reflected parts of the wave reach the receiving antenna and interfere with the reception of the wave. Radio waves can penetrate nonconducting materials, such as wood, bricks, and concrete, fairly well. They cannot pass through electrical conductors, such as water or metals. Above $v = 40$ MHz, radio waves from deep space can penetrate Earth's atmosphere. This makes radio-astronomy observations with ground-based telescopes possible.

Whenever transmission of electromagnetic energy from one location to another is required with minimal energy loss and disturbance, the waves are confined to a limited region by means of wires, coaxial cables, and, in the microwave region, waveguides. Unguided or wireless transmission is naturally preferred when the locations of receivers are unspecified or too numerous, as in the case of radio and television communications. Cable television, as the name implies, is an exception. In this case electromagnetic radiation is transmitted by a coaxial cable system to users either from a community antenna or directly from broadcasting stations. The shielding of this guided transmission from disturbances provides high-quality signals.

Figure 6 shows the electric field E (solid lines) and the magnetic field B (dashed lines) of an electromagnetic wave guided by a coaxial cable. There is a potential difference between the inner and outer conductors and so electric field lines E extend from one conductor to the other, represented here in cross section. The conductors carry opposite currents that produce the magnetic field lines B. The electric and magnetic fields are perpendicular to each other and perpendicular to the direction of propagation, as is characteristic of the electromagnetic waves illustrated in Figure 2. At any cross section viewed, the directions of the E and B field lines change to their opposite with the frequency v of the radiation. This direction reversal of the fields does not change the direction of propagation along the conductors. The speed of propagation is again the universal speed of light if the region between the conductors consists of air or free space.

A combination of radio waves and strong magnetic fields is used by magnetic resonance imaging (MRI) to produce diagnostic pictures of parts of the human body and brain without apparent harmful effects. This imaging technique has thus found increasingly wider application in medicine (see also radiation).

Extremely low-frequency (ELF) waves are of interest for communications systems for submarines. The relatively weak absorption by seawater of electromagnetic radiation at low frequencies and the existence of prominent resonances of the natural cavity formed by Earth and the ionosphere make the range between 5 and 100 Hz attractive for this application.

Microwaves

The microwave region extends from 1,000 to 300,000 MHz (or 30 cm to 1 mm wavelength). Although microwaves were first produced and studied in 1886 by Hertz, their practical application had to await the invention of suitable generators, such as the klystron and magnetron.

Microwaves are the principal carriers of high-speed data transmissions between stations on Earth and also between ground-based stations and satellites and space probes. A system of synchronous satellites about 36,000 km above Earth is used for international broadband of all kinds of communications—e.g., television and telephone.

Microwave transmitters and receivers are parabolic dish antennas. They produce microwave beams whose spreading angle is proportional to the ratio of the wavelength of the constituent waves to the diameter of the dish. The beams can thus be directed like a searchlight. Radar beams consist of short pulses of microwaves. One can determine the distance of an airplane or ship by measuring the time it takes such a pulse to travel to the object and, after reflection, back to the radar dish antenna. Moreover, by making use of the change in frequency of the reflected wave pulse caused by the Doppler effect (see above Speed of electromagnetic radiation and the Doppler effect), one can measure the speed of objects. Microwave radar is therefore widely used for guiding airplanes and vessels and for detecting speeding motorists. Microwaves can penetrate clouds of smoke but are scattered by water droplets, so they are used for mapping meteorologic disturbances and in weather forecasting.

Microwaves play an increasingly wide role in heating and cooking food. They are absorbed by water and fat in foodstuffs (e.g., in the tissue of meats) and produce heat from the inside. In most cases, this reduces the cooking time a hundredfold. Such dry objects as glass and ceramics, on the other hand, are not heated in the process, and metal foils are not penetrated at all.

The heating effect of microwaves destroys living tissue when the temperature of the tissue exceeds 43° C (109° F). Accordingly, exposure to intense microwaves in excess of 20 milliwatts of power per square centimetre of body surface is harmful. The lens of the human eye is particularly affected by waves with a frequency of 3000 MHz, and repeated and extended exposure can result in cataracts. Radio waves and microwaves of far less power (microwatts per square centimetre) than the 10–20 milliwatts per square centimetre needed to produce heating in living tissue can have adverse effects on the electrochemical balance of the brain and the development of a fetus if these waves are modulated or pulsed at low frequencies between 5 and 100 hertz, which are of the same magnitude as brain wave frequencies.

Various types of microwave generators and amplifiers have been developed. Vacuum-tube devices, the klystron and the magnetron, continue to be used on a wide scale, especially for higher-power applications. Klystrons are primarily employed as amplifiers in radio relay systems and for dielectric heating, while magnetrons have been adopted for radar systems and microwave ovens. (For a detailed discussion of these devices, see electron tube.) Solid-state technology has yielded several devices capable of producing, amplifying, detecting, and controlling microwaves. Notable among these are the Gunn diode and the tunnel (or Esaki) diode. Another type of device, the maser (acronym for "microwave amplification by stimulated emission of radiation") has proved useful in such areas as radio astronomy, microwave radiometry, and long-distance communications.

Astronomers have discovered what appears to be natural masers in some interstellar clouds. Observations of radio radiation from interstellar hydrogen (H2) and certain other molecules indicate amplification by the maser process. Also, as was mentioned above, microwave cosmic background radiation has been detected and is considered by many to be the remnant of the primeval fireball postulated by the big-bang cosmological model.

Infrared radiation

Beyond the red end of the visible range but at frequencies higher than those of radar waves and microwaves is the infrared region of the electromagnetic spectrum, between frequencies of 1012 and 5 × 1014 Hz (or wavelengths from 0.1 to 7.5 × 10–5 cm). William Herschel, a German-born British musician and self-taught astronomer, discovered this form of radiation in 1800 by exploring, with the aid of a thermometer, sunlight dispersed into its colours by a glass prism. Infrared radiation is absorbed and emitted by the rotations and vibrations of chemically bonded atoms or groups of atoms and thus by many kinds of materials. For instance, window glass that is transparent to visible light absorbs infrared radiation by the vibration of its constituent atoms. Infrared radiation is strongly absorbed by water, as shown in Figure 3, and by the atmosphere. Although invisible to the eye, infrared radiation can be detected as warmth by the skin. Nearly 50 percent of the Sun's radiant energy is emitted in the infrared region of the electromagnetic spectrum, with the rest primarily in the visible region.

Atmospheric haze and certain pollutants that scatter visible light are nearly transparent to parts of the infrared spectrum because the scattering efficiency increases with the fourth power of the frequency. Infrared photography of distant objects from the air takes advantage of this phenomenon. For the same reason, infrared astronomy enables researchers to observe cosmic objects through large clouds of interstellar dust that scatter infrared radiation substantially less than visible light. However, since water vapour, ozone, and carbon dioxide in the atmosphere absorb large parts of the infrared spectrum, many infrared astronomical observations are carried out at high altitude by balloons, rockets, aircraft, or spacecraft.

An infrared photograph of a landscape enhances objects according to their heat emission: blue sky and water appear nearly black, whereas green foliage and unexposed skin show up brightly. Infrared photography can reveal pathological tissue growths (thermography) and defects in electronic systems and circuits due to their increased emission of heat.

The infrared absorption and emission characteristics of molecules and materials yield important information about the size, shape, and chemical bonding of molecules and of atoms and ions in solids. The energies of rotation and vibration are quantized in all systems. The infrared radiation energy hν emitted or absorbed by a given molecule or substance is therefore a measure of the difference of some of the internal energy states. These in turn are determined by the atomic weight and molecular bonding forces. For this reason, infrared spectroscopy is a powerful tool for determining the internal structure of molecules and substances or, when such information is already known and tabulated, for identifying the amounts of those species in a given sample. Infrared spectroscopic techniques are often used to determine the composition and hence the origin and age of archaeological specimens and for detecting forgeries of art and other objects, which, when inspected under visible light, resemble the originals.

Infrared radiation plays an important role in heat transfer and is integral to the so-called greenhouse effect (see above The greenhouse effect of the atmosphere), influencing the thermal radiation budget of Earth on a global scale and affecting nearly all biospheric activity. Virtually every object at Earth's surface emits electromagnetic radiation primarily in the infrared region of the spectrum.

Artificial sources of infrared radiation include, besides hot objects, infrared light-emitting diodes (LEDs) and lasers. LEDs are small inexpensive optoelectronic devices made of such semiconducting materials as gallium arsenide. Infrared LEDs are employed as optoisolators and as light sources in some fibre-optics-based communications systems. Powerful optically pumped infrared lasers have been developed by using carbon dioxide and carbon monoxide. Carbon dioxide infrared lasers are used to induce and alter chemical reactions and in isotope separation. They also are employed in lidar systems. Other applications of infrared light include its use in the range finders of automatic self-focusing cameras, security alarm systems, and night-vision optical instruments.

Instruments for detecting infrared radiation include heat-sensitive devices such as thermocouple detectors, bolometers (some of these are cooled to temperatures close to absolute zero so that the thermal radiation of the detector system itself is greatly reduced), photovoltaic cells, and photoconductors. The latter are made of semiconductor materials (e.g., silicon and lead sulfide) whose electrical conductance increases when exposed to infrared radiation.

Visible radiation

Visible light is the most familiar form of electromagnetic radiation and makes up that portion of the spectrum to which the eye is sensitive. This span is very narrow; the frequencies of violet light are only about twice those of red. The corresponding wavelengths extend from 7×10^{-5} cm (red) to 4×10^{-5} cm (violet). The energy of a photon from the centre of the visible spectrum (yellow) is $h\nu = 2.2$ eV. This is one million times larger than the energy of a photon of a television wave and one billion times larger than that of radio waves in general (see Figure 1).

Life on Earth could not exist without visible light, which represents the peak of the Sun's spectrum and close to one-half of all of its radiant energy. Visible light is essential for photosynthesis, which enables plants to produce the carbohydrates and proteins that are the food sources for animals. Coal and oil are sources of energy accumulated from sunlight in plants and microorganisms millions of years ago, and hydroelectric power is extracted from one step of the hydrologic cycle kept in motion by sunlight at the present time.

Considering the importance of visible sunlight for all aspects of terrestrial life, one cannot help being awed by the absorption spectrum of water in Figure 3. The remarkable transparency of water centred in the narrow regime of visible light, indicated by vertical dashed lines in Figure 3, is the result of the characteristic distribution of internal energy states of water. Absorption is strong toward the infrared on account of molecular vibrations and intermolecular oscillations. In the ultraviolet region, absorption of radiation is caused by electronic excitations. Light of frequencies having absorption coefficients larger than $\alpha = 10$ cm^{-1} cannot even reach the retina of the human eye, because its constituent liquid consists mainly of water that absorbs such frequencies of light.

Since the 1970s an increasing number of devices have been developed for converting sunlight into electricity. Unlike various conventional energy sources, solar energy does not become depleted by use and does not pollute the environment. Two branches of development may be noted—namely, photothermal and photovoltaic technologies. In photothermal devices, sunlight is used to heat a substance, as, for example, water, to produce steam with which to drive a generator. Photovoltaic devices, on the other hand, convert the energy in sunlight directly to electricity by use of the photovoltaic effect in a semiconductor junction. Solar panels consisting of photovoltaic devices made of gallium arsenide have conversion efficiencies of more than 20 percent and are used to provide electric power in many satellites and space probes. Solar cells have replaced dry-cell batteries in some portable electronic instruments, and solar energy power stations of more than 500 megawatts capacity have been built.

The intensity and spectral composition of visible light can be measured and recorded by essentially any process or property that is affected by light. Detectors make use of a photographic process based on silver halide, the photoemission of electrons from metal surfaces, the generation of electric current in a photovoltaic cell, and the increase in electrical conduction in semiconductors.

Glass fibres constitute an effective means of guiding and transmitting light. A beam of light is confined by total internal reflection to travel inside such an optical fibre, whose thickness may be anywhere between one hundredth of a millimetre and a few millimetres. Many thin optical fibres can be combined into bundles to achieve image reproduction. The flexibility of these fibres or fibre bundles permits their use in medicine for optical exploration of internal organs. Optical fibres connecting the continents provide the capability to transmit substantially larger amounts of information than other systems of international telecommunications. Another advantage of optical fibre communication systems is that transmissions cannot easily be intercepted and are not disturbed by lower atmospheric and stratospheric disturbances.

Optical fibres integrated with miniature semiconductor lasers and light-emitting diodes, as well as with light detector arrays and photoelectronic imaging and recording materials, form the building blocks of a new optoelectronics industry. Some familiar commercial products are optoelectronic copying machines, laser printers, compact disc players, optical recording media, and optical disc mass-storage systems of exceedingly high bit density.

Ultraviolet radiation

The German physicist Johann Wilhelm Ritter, having learned of Herschel's discovery of infrared waves, looked beyond the violet end of the visible spectrum of the Sun and found (in 1801) that there exist invisible rays that darken silver chloride even more efficiently than visible light. This spectral region extending between visible light and X-rays is designated ultraviolet. Sources of this form of electromagnetic radiation are hot objects like the Sun, synchrotron radiation sources, mercury or xenon arc lamps, and gaseous discharge tubes filled with gas atoms (e.g., mercury, deuterium, or hydrogen) that have internal electron energy levels which correspond to the photons of ultraviolet light.

When ultraviolet light strikes certain materials, it causes them to fluoresce—i.e., they emit electromagnetic radiation of lower energy, such as visible light. The spectrum of fluorescent light is characteristic of a material's composition and thus can be used for screening minerals, detecting bacteria in spoiled food, identifying pigments, or detecting forgeries of artworks and other objects (the aged surfaces of ancient marble sculptures, for instance, fluoresce yellow-green, whereas a freshly cut marble surface fluoresces bright violet).

Optical instruments for the ultraviolet region are made of special materials, such as quartz, certain silicates, and metal fluorides, which are transparent at least in the near ultraviolet. Far-ultraviolet radiation is absorbed by nearly all gases and materials and thus requires reflection optics in vacuum chambers.

Ultraviolet radiation is detected by photographic plates and by means of the photoelectric effect in photomultiplier tubes. Also, ultraviolet radiation can be converted to visible light by fluorescence before detection.

The relatively high energy of ultraviolet light gives rise to certain photochemical reactions. This characteristic is exploited to produce cyanotype impressions on fabrics and for blueprinting design drawings. Here, the fabric or paper is treated with a mixture of chemicals that react upon exposure to ultraviolet light to form an insoluble blue compound. Electronic excitations caused by ultraviolet radiation also produce changes in the colour and transparency of photosensitive and photochromic glasses. Photochemical and photostructural changes in certain polymers constitute the basis for photolithography and the processing of the microelectronic circuits.

Although invisible to the eyes of humans and most vertebrates, near-ultraviolet light can be seen by many insects. Butterflies and many flowers that appear to have identical colour patterns under visible light are distinctly different when viewed under the ultraviolet rays perceptible to insects.

An important difference between ultraviolet light and electromagnetic radiation of lower frequencies is the ability of the former to ionize, meaning that it can knock an electron out from atoms and molecules. All high-frequency electromagnetic radiation beyond the visible—i.e., ultraviolet light, X-rays, and gamma rays—is ionizing and therefore harmful to body tissues, living cells, and DNA (deoxyribonucleic acid). The harmful effects of ultraviolet light to humans and larger animals are mitigated by the fact that this form of radiation does not penetrate much further than the skin.

The body of a sunbather is struck by 1021 photons every second, and 1 percent of these, or more than a billion billion per second, are photons of ultraviolet radiation. Tanning and natural body pigments help to protect the skin to some degree, preventing the destruction of skin cells by ultraviolet light. Nevertheless, overexposure to the ultraviolet component of sunlight can cause skin cancer, cataracts of the eyes, and damage to the body's immune system. Fortunately, a layer of ozone ($O3$) in the stratosphere absorbs the most-damaging ultraviolet rays, which have wavelengths of 2000 and 2900 angstroms (one angstrom [Å] = 10–10 metre), and attenuates those with wavelengths between 2900 and 3150 Å. Without this protective layer of ozone, life on Earth would not be possible. The ozone layer is produced at an altitude of about 10 to 50 km (6 to 30 miles) above Earth's surface by a reaction between upward-diffusing molecular oxygen ($O2$) and downward-diffusing ionized atomic oxygen ($O+$). In the late 20th century this life-protecting stratospheric ozone layer was reduced by chlorine atoms in chlorofluorocarbon (or Freon) gases released into the atmosphere by aerosol propellants, air-conditioner coolants, solvents used in the manufacture of electronic components, and other sources. Limits were placed on the sale of ozone-depleting chemicals, and the ozone layer was expected to recover eventually.

Ionized atomic oxygen, nitrogen, and nitric oxide are produced in the upper atmosphere by absorption of solar ultraviolet radiation. This ionized region is the ionosphere, which affects radio communications and reflects and absorbs radio waves of frequencies below 40 MHz.

X-rays

The German physicist Wilhelm Conrad Röntgen discovered X-rays in 1895 by accident while studying cathode rays in a low-pressure gas discharge tube. (A few years later J.J. Thomson of England showed that cathode rays were electrons emitted from the negative electrode [cathode] of the discharge tube.) Röntgen noticed the fluorescence of a barium platinocyanide screen that happened to lie near the discharge tube. He traced the source of the hitherto undetected form of radiation to the point where the cathode rays hit the wall of the discharge tube, and he mistakenly concluded from his inability to observe reflection or refraction that his new rays were unrelated to light. Because of his uncertainty about their nature, he called them X-radiation. This early failure can be attributed to the very short wavelengths of X-rays (10–8 to 10–11 cm), which correspond to photon energies from 200 to 100,000 eV. In 1912 another German physicist, Max von Laue, realized that the regular arrangement of atoms in crystals should provide a natural grating of the right spacing (about 10–8 cm) to produce an interference pattern on a photographic plate when X-rays pass through such a crystal. The success of this experiment, carried out by Walter Friedrich and Paul Knipping, not only identified X-rays with electromagnetic radiation but also initiated the use of X-rays for studying the detailed atomic structure of crystals. The interference of X-rays diffracted in certain directions from crystals in so-called X-ray diffractometers, in turn, permits the dissection of X-rays into their different frequencies, just as a prism diffracts and spreads the various colours of light. The spectral composition and characteristic frequencies of X-rays emitted by a given X-ray source can thus be measured. As in optical spectroscopy, the X-ray photons emitted correspond to the differences of the internal electronic energies in atoms and molecules. Because of their much higher energies, however, X-ray photons are associated with the inner-shell electrons close to the atomic nuclei, whereas optical absorption and emission are related to the outermost electrons in atoms or in materials in general. Since the outer electrons are used for chemical bonding while the energies of inner-shell electrons remain essentially unaffected by atomic bonding, the identity and quantity of elements that make up a material are more accurately determined by the emission, absorption, or fluorescence of X-rays than of photons of visible or ultraviolet light.

The contrast between body parts in medical X-ray photographs (radiographs) is produced by the different scattering and absorption of X-rays by bones and tissues. Within months of Röntgen's discovery of X-rays and his first X-ray photograph of his wife's hand, this form of electromagnetic radiation became indispensable in orthopedic and dental medicine. The use of X-rays for obtaining images of the body's interior has undergone considerable development over the years and has culminated in the highly sophisticated procedure known as computed tomography (CAT; see radiation).

Notwithstanding their usefulness in medical diagnosis, the ability of X-rays to ionize atoms and molecules and their penetrating power make them a potential health hazard. Exposure of body cells and tissue to large doses of such ionizing radiation can result in abnormalities in DNA that may lead to cancer and birth defects. (For a detailed treatment of the effects of X-rays and other forms of ionizing radiation on human health and the levels of such radiation encountered in daily life, see radiation: Biological effects of ionizing radiation.)

X-rays are produced in X-ray tubes by the deceleration of energetic electrons (bremsstrahlung) as they hit a metal target or by accelerating electrons moving at relativistic velocities in circular orbits (synchrotron radiation; see above Continuous spectra of electromagnetic radiation). They are detected by their photochemical action in photographic emulsions or by their ability to ionize gas atoms. Every X-ray photon produces a burst of electrons and ions, resulting in a current pulse. By counting the rate of such current pulses per second, the intensity of a flux of X-rays can be measured. Instruments used for this purpose are called Geiger counters.

X-ray astronomy has revealed very strong sources of X-rays in deep space. In the Milky Way Galaxy, of which the solar system is a part, the most-intense sources are certain double-star systems in which one of the two stars is thought to be either a compact neutron star or a black hole. The ionized gas of the circling companion star falls by gravitation into the compact star, generating X-rays that may be more than 1,000 times as intense as the total amount of light emitted by the Sun. At the moment of their explosion, supernovae emit a good fraction of their energy in a burst of X-rays.

Gamma rays

Six years after the discovery of radioactivity (1896) by Henri Becquerel of France, the New Zealand-born British physicist Ernest Rutherford found that three different kinds of radiation are emitted in the decay of radioactive substances; these he called alpha, beta, and gamma rays in sequence of their ability to penetrate matter. The alpha particles were found to be identical with the nuclei of helium atoms, and the beta rays were identified as electrons. In 1912 it was shown that the much more penetrating gamma rays have all the properties of very energetic electromagnetic radiation, or photons. Gamma-ray photons are between 10,000 and 10,000,000 times more energetic than the photons of visible light when they originate from radioactive atomic nuclei. Gamma rays with a million million times higher energy make up a very small part of the cosmic rays that reach Earth from supernovae or from other galaxies. The origin of the most-energetic gamma rays is not yet known.

During radioactive decay, an unstable nucleus usually emits alpha particles, electrons, gamma rays, and neutrinos spontaneously. In nuclear fission, the unstable nucleus breaks into fragments, which are themselves complex nuclei, along with such particles as neutrons and protons. The resultant stable nuclei or nuclear fragments are usually in a highly excited state and then reach their low-energy ground state by emitting one or more gamma rays. Such a decay scheme is shown schematically in Figure 7 for the unstable nucleus sodium-24 (24Na). Much of what is known about the internal structure and energies of nuclei has been obtained from the emission or resonant absorption of gamma rays by nuclei. Absorption of gamma rays by nuclei can cause them to eject neutrons or alpha particles or it can even split a nucleus like a bursting bubble in what is called photodisintegration. A gamma particle hitting a hydrogen nucleus (that is, a proton), for example, produces a positive pi-meson and a neutron or a neutral pi-meson and a proton. Neutral pi-mesons, in turn, have a very brief mean life of $1.8 \times 10{-}16$ second and decay into two gamma rays of energy $h\nu \approx 70$ MeV. When an energetic gamma ray $h\nu > 1.02$ MeV passes a nucleus, it may disappear while creating an electron–positron pair. Gamma photons interact with matter by discrete elementary processes that include resonant absorption, photodisintegration, ionization, scattering (Compton scattering), or pair production.

Gamma rays are detected by their ability to ionize gas atoms or to create electron–hole pairs in semiconductors or insulators. By counting the rate of charge pulses or voltage pulses or by measuring the scintillation of the light emitted by the subsequently recombining electron–hole pairs, one can determine the number and energy of gamma rays striking an ionization detector or scintillation counter.

Both the specific energy of the gamma-ray photon emitted as well as the half-life of the specific radioactive decay process that yields the photon identify the type of nuclei at hand and their concentrations. By bombarding stable nuclei with neutrons, one can artificially convert more than 70 different stable nuclei into radioactive nuclei and use their characteristic gamma emission for purposes of identification, for impurity analysis of metallurgical specimens (neutron-activation analysis), or as radioactive tracers with which to determine the functions or malfunctions of human organs, to follow the life cycles of organisms, or to determine the effects of chemicals on biological systems and plants.

The great penetrating power of gamma rays stems from the fact that they have no electric charge and thus do not interact with matter as strongly as do charged particles. Because of their penetrating power gamma rays can be used for radiographing holes and defects in metal castings and other structural parts. At the same time, this property makes gamma rays extremely hazardous. The lethal effect of this form of ionizing radiation makes it useful for sterilizing medical supplies that cannot be sanitized by boiling or for killing organisms that cause food spoilage. More than 50 percent of the ionizing radiation to which humans are exposed comes from natural radon gas, which is an end product of the radioactive decay chain of natural radioactive substances in minerals. Radon escapes from the ground and enters the environment in varying amounts.

9

Data Mining

---·🎵·---

Data mining, also called knowledge discovery in databases, in computer science, the process of discovering interesting and useful patterns and relationships in large volumes of data. The field combines tools from statistics and artificial intelligence (such as neural networks and machine learning) with database management to analyze large digital collections, known as data sets. Data mining is widely used in business (insurance, banking, retail), science research (astronomy, medicine), and government security (detection of criminals and terrorists).

The proliferation of numerous large, and sometimes connected, government and private databases has led to regulations to ensure that individual records are accurate and secure from unauthorized viewing or tampering. Most types of data mining are targeted toward ascertaining general knowledge about a group rather than knowledge about specific individuals—a supermarket is less concerned about selling one more item to one person than about selling many items to many people—though pattern analysis also may be used to discern anomalous individual behaviour such as fraud or other criminal activity.

Origins and early applications

As computer storage capacities increased during the 1980s, many companies began to store more transactional data. The resulting record collections, often called data warehouses, were too large to be analyzed with traditional statistical approaches. Several computer science conferences and workshops were held to consider how recent advances in the field of artificial intelligence (AI)—such as discoveries from expert systems, genetic algorithms, machine learning, and neural networks—could be adapted for knowledge discovery (the preferred term in the computer science community). The process led in 1995 to the First International Conference on Knowledge Discovery and Data Mining, held in Montreal, and the launch in 1997 of the journal Data Mining and Knowledge Discovery. This was also the period when many early data-mining companies were formed and products were introduced.

One of the earliest successful applications of data mining, perhaps second only to marketing research, was credit-card-fraud detection. By studying a consumer's purchasing behaviour, a typical pattern usually becomes apparent; purchases made outside this pattern can then be flagged for later investigation or to deny a transaction. However, the wide variety of normal behaviours makes this challenging; no single distinction between normal and fraudulent behaviour works for everyone or all the time. Every individual is likely to make some purchases that differ from the types he has made before, so relying on what is normal for a single individual is likely to give too many false alarms. One approach to improving reliability is first to group individuals that have similar purchasing patterns, since group models are less sensitive to minor anomalies. For example, a "frequent business travelers" group will likely have a pattern that includes unprecedented purchases in diverse locations, but members of this group might be flagged for other transactions, such as catalog purchases, that do not fit that group's profile.

Modeling and data-mining approaches

Model creation

The complete data-mining process involves multiple steps, from understanding the goals of a project and what data are available to implementing process changes based on the final analysis. The three key computational steps are the model-learning process, model evaluation, and use of the model. This division is clearest with classification of data. Model learning occurs when one algorithm is applied to data about which the group (or class) attribute is known in order to produce a classifier, or an algorithm learned from the data. The classifier is then tested with an independent evaluation set that contains data with known attributes. The extent to which the model's classifications agree with the known class for the target attribute can then be used to determine the expected accuracy of the model. If the model is sufficiently accurate, it can be used to classify data for which the target attribute is unknown.

Data-mining techniques

There are many types of data mining, typically divided by the kind of information (attributes) known and the type of knowledge sought from the data-mining model.

Predictive modeling

Predictive modeling is used when the goal is to estimate the value of a particular target attribute and there exist sample training data for which values of that attribute are known. An example is classification, which takes a set of data already divided into predefined groups and searches for patterns in the data that differentiate those groups. These discovered patterns then can be used to classify other data where the right group designation for the target attribute is unknown (though other attributes may be known). For instance, a manufacturer could develop a predictive model that distinguishes parts that fail under extreme heat, extreme cold, or other conditions based on their manufacturing environment, and this model may then be used to determine appropriate applications for each part. Another technique employed in predictive modeling is regression analysis, which can be used when the target attribute is a numeric value and the goal is to predict that value for new data.

Descriptive modeling

Descriptive modeling, or clustering, also divides data into groups. With clustering, however, the proper groups are not known in advance; the patterns discovered by analyzing the data are used to determine the groups. For example, an advertiser could analyze a general population in order to classify potential customers into different clusters and then develop separate advertising campaigns targeted to each group. Fraud detection also makes use of clustering to identify groups of individuals with similar purchasing patterns.

Pattern mining

Pattern mining concentrates on identifying rules that describe specific patterns within the data. Market-basket analysis, which identifies items that typically occur together in purchase transactions, was one of the first applications of data mining. For example, supermarkets used market-basket analysis to identify items that were often purchased together—for instance, a store featuring a fish sale would also stock up on tartar sauce. Although testing for such associations has long been feasible and is often simple to see in small data sets, data mining has enabled the discovery of less apparent associations in immense data sets. Of most interest is the discovery of unexpected associations, which may open new avenues for marketing or research. Another important use of pattern mining is the discovery of sequential patterns; for example, sequences of errors or warnings that precede an equipment failure may be used to schedule preventative maintenance or may provide insight into a design flaw.

Anomaly detection

Anomaly detection can be viewed as the flip side of clustering—that is, finding data instances that are unusual and do not fit any established pattern. Fraud detection is an example of anomaly detection. Although fraud detection may be viewed as a problem for predictive modeling, the relative rarity of fraudulent transactions and the speed with which criminals develop new types of fraud mean that any predictive model is likely to be of low accuracy and to quickly become out of date. Thus, anomaly detection instead concentrates on modeling what is normal behaviour in order to identify unusual transactions. Anomaly detection also is used with various monitoring systems, such as for intrusion detection.

Numerous other data-mining techniques have been developed, including pattern discovery in time series data (e.g., stock prices), streaming data (e.g., sensor networks), and relational learning (e.g., social networks).

Privacy concerns and future directions

The potential for invasion of privacy using data mining has been a concern for many people. Commercial databases may contain detailed records of people's medical history, purchase transactions, and telephone usage, among other aspects of their lives. Civil libertarians consider some databases held by businesses and governments to be an unwarranted intrusion and an invitation to abuse. For example, the American Civil Liberties Union sued the U.S. National Security Agency (NSA) alleging warrantless spying on American citizens through the acquisition of call records from some American telecommunication companies. The program, which began in 2001, was not discovered by the public until 2006, when the information began to leak out. Often the risk is not from data mining itself (which usually aims to produce general knowledge rather than to learn information about specific issues) but from misuse or inappropriate disclosure of information in these databases.

In the United States, many federal agencies are now required to produce annual reports that specifically address the privacy implications of their data-mining projects. The U.S. law requiring privacy reports from federal agencies defines data mining quite restrictively as "...analyses to discover or locate a predictive pattern or anomaly indicative of terrorist or criminal activity on the part of any individual or individuals." As various local, national, and international law-enforcement agencies have begun to share or integrate their databases, the potential for abuse or security breaches has forced governments to work with industry on developing more secure computers and networks. In particular, there has been research in techniques for privacy-preserving data mining that operate on distorted, transformed, or encrypted data to decrease the risk of disclosure of any individual's data.

Data mining is evolving, with one driver being competitions on challenge problems. A commercial example of this was the $1 million Netflix Prize. Netflix, an American company that offers movie rentals delivered by mail or streamed over the Internet, began the contest in 2006 to see if anyone could improve by 10 percent its recommendation system, an algorithm for predicting an individual's movie preferences based on previous rental data. The prize was awarded on Sept. 21, 2009, to BellKor's Pragmatic Chaos—a team of seven mathematicians, computer scientists, and engineers from the United States, Canada, Austria, and Israel who had achieved the 10 percent goal on June 26, 2009, and finalized their victory with an improved algorithm 30 days later. The three-year open competition had spurred many clever data-mining innovations from contestants. For example, the 2007 and 2008 Conferences on Knowledge Discovery and Data Mining held workshops on the Netflix Prize, at which research papers were presented on topics ranging from new collaborative filtering techniques to faster matrix factorization (a key component of many recommendation systems). Concerns over privacy of such data have also led to advances in understanding privacy and anonymity.

Data mining is not a panacea, however, and results must be viewed with the same care as with any statistical analysis. One of the strengths of data mining is the ability to analyze quantities of data that would be impractical to analyze manually, and the patterns found may be complex and difficult for humans to understand; this complexity requires care in evaluating the patterns. Nevertheless, statistical evaluation techniques can result in knowledge that is free from human bias, and the large amount of data can reduce biases inherent in smaller samples. Used properly, data mining provides valuable insights into large data sets that otherwise would not be practical or possible to obtain.

10
Biofuel

—•ᑭ•—

Biofuel, any fuel that is derived from biomass—that is, plant or algae material or animal waste. Since such feedstock material can be replenished readily, biofuel is considered to be a source of renewable energy, unlike fossil fuels such as petroleum, coal, and natural gas. Biofuel is commonly advocated as a cost-effective and environmentally benign alternative to petroleum and other fossil fuels, particularly within the context of rising petroleum prices and increased concern over the contributions made by fossil fuels to global warming. Many critics express concerns about the scope of the expansion of certain biofuels because of the economic and environmental costs associated with the refining process and the potential removal of vast areas of arable land from food production.

Types of biofuels

Some long-exploited biofuels, such as wood, can be used directly as a raw material that is burned to produce heat. The heat, in turn, can be used to run generators in a power plant to produce electricity. A number of existing power facilities burn grass, wood, or other kinds of biomass.

Liquid biofuels are of particular interest because of the vast infrastructure already in place to use them, especially for transportation. The liquid biofuel in greatest production is ethanol (ethyl alcohol), which is made by fermenting starch or sugar. Brazil and the United States are among the leading producers of ethanol. In the United States ethanol biofuel is made primarily from corn (maize) grain, and it is typically blended with gasoline to produce "gasohol," a fuel that is 10 percent ethanol. In Brazil, ethanol biofuel is made primarily from sugarcane, and it is commonly used as a 100-percent-ethanol fuel or in gasoline blends containing 85 percent ethanol. Unlike the "first-generation" ethanol biofuel produced from food crops, "second-generation" cellulosic ethanol is derived from low-value biomass that possesses a high cellulose content, including wood chips, crop residues, and municipal waste. Cellulosic ethanol is commonly made from sugarcane bagasse, a waste product from sugar processing, or from various grasses that can be cultivated on low-quality land. Given that the conversion rate is lower than with first-generation biofuels, cellulosic ethanol is dominantly used as a gasoline additive.

The second most common liquid biofuel is biodiesel, which is made primarily from oily plants (such as the soybean or oil palm) and to a lesser extent from other oily sources (such as waste cooking fat from restaurant deep-frying). Biodiesel, which has found greatest acceptance in Europe, is used in diesel engines and usually blended with petroleum diesel fuel in various percentages. The use of algae and cyanobacteria as a source of "third-generation" biodiesel holds promise but has been difficult to develop economically. Some algal species contain up to 40 percent lipids by weight, which can be converted into biodiesel or synthetic petroleum. Some estimates state that algae and cyanobacteria could yield between 10 and 100 times more fuel per unit area than second-generation biofuels.

Other biofuels include methane gas and biogas—which can be derived from the decomposition of biomass in the absence of oxygen—and methanol, butanol, and dimethyl ether—which are in development.

Economic and environmental considerations

In evaluating the economic benefits of biofuels, the energy required to produce them has to be taken into account. For example, the process of growing corn to produce ethanol consumes fossil fuels in farming equipment, in fertilizer manufacturing, in corn transportation, and in ethanol distillation. In this respect, ethanol made from corn represents a relatively small energy gain; the energy gain from sugarcane is greater and that from cellulosic ethanol or algae biodiesel could be even greater.

Biofuels also supply environmental benefits but, depending on how they are manufactured, can also have serious environmental drawbacks. As a renewable energy source, plant-based biofuels in principle make little net contribution to global warming and climate change; the carbon dioxide (a major greenhouse gas) that enters the air during combustion will have been removed from the air earlier as growing plants engage in photosynthesis. Such a material is said to be "carbon neutral." In practice, however, the industrial production of agricultural biofuels can result in additional emissions of greenhouse gases that may offset the benefits of using a renewable fuel. These emissions include carbon dioxide from the burning of fossil fuels during the production process and nitrous oxide from soil that has been treated with nitrogen fertilizer. In this regard, cellulosic biomass is considered to be more beneficial.

Land use is also a major factor in evaluating the benefits of biofuels. The use of regular feedstock, such as corn and soybeans, as a primary component of first-generation biofuels sparked the "food versus fuel" debate. In diverting arable land and feedstock from the human food chain, biofuel production can affect the economics of food price and availability. In addition, energy crops grown for biofuel can compete for the world's natural habitats. For example, emphasis on ethanol derived from corn is shifting grasslands and brushlands to corn monocultures, and emphasis on biodiesel is bringing down ancient tropical forests to make way for oil palm plantations. Loss of natural habitat can change the hydrology, increase erosion, and generally reduce biodiversity of wildlife areas. The clearing of land can also result in the sudden release of a large amount of carbon dioxide as the plant matter that it contains is burned or allowed to decay.

Some of the disadvantages of biofuels apply mainly to low-diversity biofuel sources—corn, soybeans, sugarcane, oil palms—which are traditional agricultural crops. One alternative involves the use of highly diverse mixtures of species, with the North American tallgrass prairie as a specific example. Converting degraded agricultural land that is out of production to such high-diversity biofuel sources could increase wildlife area, reduce erosion, cleanse waterborne pollutants, store carbon dioxide from the air as carbon compounds in the soil, and ultimately restore fertility to degraded lands. Such biofuels could be burned directly to generate electricity or converted to liquid fuels as technologies develop.

The proper way to grow biofuels to serve all needs simultaneously will continue to be a matter of much experimentation and debate, but the fast growth in biofuel production will likely continue. In the United States the Energy Independence and Security Act of 2007 mandated the use of 136 billion litres (36 billion gallons) of biofuels annually by 2022, more than a sixfold increase over 2006 production levels. The legislation also requires, with certain stipulations, that 79 billion litres (21 billion gallons) of the total amount be biofuels other than corn-derived ethanol, and it continued certain government subsidies and tax incentives for biofuel production.

One distinctive promise of biofuels is that, in combination with an emerging technology called carbon capture and storage, the process of producing and using biofuels may be capable of perpetually removing carbon dioxide from the atmosphere. Under this vision, biofuel crops would remove carbon dioxide from the air as they grow, and energy facilities would capture the carbon dioxide given off as biofuels are burned to generate power. Captured carbon dioxide could be sequestered (stored) in long-term repositories such as geologic formations beneath the land, in sediments of the deep ocean, or conceivably as solids such as carbonates.

11
Thermodynamics

————•⎘•————

Thermodynamics, science of the relationship between heat, work, temperature, and energy. In broad terms, thermodynamics deals with the transfer of energy from one place to another and from one form to another. The key concept is that heat is a form of energy corresponding to a definite amount of mechanical work.

Heat was not formally recognized as a form of energy until about 1798, when Count Rumford (Sir Benjamin Thompson), a British military engineer, noticed that limitless amounts of heat could be generated in the boring of cannon barrels and that the amount of heat generated is proportional to the work done in turning a blunt boring tool. Rumford's observation of the proportionality between heat generated and work done lies at the foundation of thermodynamics. Another pioneer was the French military engineer Sadi Carnot, who introduced the concept of the heat-engine cycle and the principle of reversibility in 1824. Carnot's work concerned the limitations on the maximum amount of work that can be obtained from a steam engine operating with a high-temperature heat transfer as its driving force. Later that century, these ideas were developed by Rudolf Clausius, a German mathematician and physicist, into the first and second laws of thermodynamics, respectively.

The most important laws of thermodynamics are:

The zeroth law of thermodynamics. When two systems are each in thermal equilibrium with a third system, the first two systems are in thermal equilibrium with each other. This property makes it meaningful to use thermometers as the "third system" and to define a temperature scale.

The first law of thermodynamics, or the law of conservation of energy. The change in a system's internal energy is equal to the difference

66

between heat added to the system from its surroundings and work done by the system on its surroundings.

The second law of thermodynamics. Heat does not flow spontaneously from a colder region to a hotter region, or, equivalently, heat at a given temperature cannot be converted entirely into work. Consequently, the entropy of a closed system, or heat energy per unit temperature, increases over time toward some maximum value. Thus, all closed systems tend toward an equilibrium state in which entropy is at a maximum and no energy is available to do useful work. This asymmetry between forward and backward processes gives rise to what is known as the "arrow of time."

The third law of thermodynamics. The entropy of a perfect crystal of an element in its most stable form tends to zero as the temperature approaches absolute zero. This allows an absolute scale for entropy to be established that, from a statistical point of view, determines the degree of randomness or disorder in a system.

Although thermodynamics developed rapidly during the 19th century in response to the need to optimize the performance of steam engines, the sweeping generality of the laws of thermodynamics makes them applicable to all physical and biological systems. In particular, the laws of thermodynamics give a complete description of all changes in the energy state of any system and its ability to perform useful work on its surroundings.

This article covers classical thermodynamics, which does not involve the consideration of individual atoms or molecules. Such concerns are the focus of the branch of thermodynamics known as statistical thermodynamics, or statistical mechanics, which expresses macroscopic thermodynamic properties in terms of the behaviour of individual particles and their interactions. It has its roots in the latter part of the 19th century, when atomic and molecular theories of matter began to be generally accepted.

Fundamental concepts

Thermodynamic states

The application of thermodynamic principles begins by defining a system that is in some sense distinct from its surroundings. For example, the system could be a sample of gas inside a cylinder with a movable piston, an entire steam engine, a marathon runner, the planet Earth, a neutron star, a black hole, or even the entire universe. In general, systems are free to exchange heat, work, and other forms of energy with their surroundings.

A system's condition at any given time is called its thermodynamic state. For a gas in a cylinder with a movable piston, the state of the system is identified by the temperature, pressure, and volume of the gas. These properties are characteristic parameters that have definite values at each state and are independent of the way in which the system arrived at that state. In other words, any change in value of a property depends only on the initial and final states of the system, not on the path followed by the system from one state to another. Such properties are called state functions. In contrast, the work done as the piston moves and the gas expands and the heat the gas absorbs from its surroundings depend on the detailed way in which the expansion occurs.

The behaviour of a complex thermodynamic system, such as Earth's atmosphere, can be understood by first applying the principles of states and properties to its component parts—in this case, water, water vapour, and the various gases making up the atmosphere. By isolating samples of material whose states and properties can be controlled and manipulated, properties and their interrelations can be studied as the system changes from state to state.

Thermodynamic equilibrium

A particularly important concept is thermodynamic equilibrium, in which there is no tendency for the state of a system to change spontaneously. For example, the gas in a cylinder with a movable piston will be at equilibrium if the temperature and pressure inside are uniform and if the restraining force on the piston is just sufficient to keep it from moving. The system can then be made to change to a new state only by an externally imposed change in one of the state functions, such as the temperature by adding heat or the volume by moving the piston. A sequence of one or more such steps connecting different states of the system is called a process. In general, a system is not in equilibrium as it adjusts to an abrupt change in its environment. For example, when a balloon bursts, the compressed gas inside is suddenly far from equilibrium, and it rapidly expands until it reaches a new equilibrium state. However, the same final state could be achieved by placing the same compressed gas in a cylinder with a movable piston and applying a sequence of many small increments in volume (and temperature), with the system being given time to come to equilibrium after each small increment. Such a process is said to be reversible because the system is at (or near) equilibrium at each step along its path, and the direction of change could be reversed at any point. This example illustrates how two different paths can connect the same initial and final states. The first is irreversible (the balloon bursts), and the second is reversible. The concept of reversible processes is something like motion without friction in mechanics. It represents an idealized limiting case that is very useful in discussing the properties of real systems. Many of the results of thermodynamics are derived from the properties of reversible processes.

Temperature

The concept of temperature is fundamental to any discussion of thermodynamics, but its precise definition is not a simple matter. For example, a steel rod feels colder than a wooden rod at room temperature simply because steel is better at conducting heat away from the skin. It is therefore necessary to have an objective way of measuring temperature. In general, when two objects are brought into thermal contact, heat will flow between them until they come into equilibrium with each other. When the flow of heat stops, they are said to be at the same temperature. The zeroth law of thermodynamics formalizes this by asserting that if an object A is in simultaneous thermal equilibrium with two other objects B and C, then B and C will be in thermal equilibrium with each other if brought into thermal contact. Object A can then play the role of a thermometer through some change in its physical properties with temperature, such as its volume or its electrical resistance.

With the definition of equality of temperature in hand, it is possible to establish a temperature scale by assigning numerical values to certain easily reproducible fixed points. For example, in the Celsius (°C) temperature scale, the freezing point of pure water is arbitrarily assigned a temperature of 0 °C and the boiling point of water the value of 100 °C (in both cases at 1 standard atmosphere; see atmospheric pressure). In the Fahrenheit (°F) temperature scale, these same two points are assigned the values 32 °F and 212 °F, respectively. There are absolute temperature scales related to the second law of thermodynamics. The absolute scale related to the Celsius scale is called the Kelvin (K) scale, and that related to the Fahrenheit scale is called the Rankine (°R) scale. These scales are related by the equations K = °C + 273.15, °R = °F + 459.67, and °R = 1.8 K. Zero in both the Kelvin and Rankine scales is at absolute zero.

Work and energy

Energy has a precise meaning in physics that does not always correspond to everyday language, and yet a precise definition is somewhat elusive. The word is derived from the Greek word ergon, meaning work, but the term work itself acquired a technical meaning with the advent of Newtonian mechanics. For example, a man pushing on a car may feel that he is doing a lot of work, but no work is actually done unless the car moves. The work done is then the product of the force applied by the man multiplied by the distance through which the car moves. If there is no friction and the surface is level, then the car, once set in motion, will continue rolling indefinitely with constant speed. The rolling car has something that a stationary car does not have—it has kinetic energy of motion equal to the work required to achieve that state of motion. The introduction of the concept of energy in this way is of great value in mechanics because, in the absence of friction, energy is never lost from the system, although it can be converted from one form to another. For example, if a coasting car comes to a hill, it will roll some distance up the hill before coming to a temporary stop. At that moment its kinetic energy of motion has been converted into its potential energy of position, which is equal to the work required to lift the car through the same vertical distance. After coming to a stop, the car will then begin rolling back down the hill until it has completely recovered its kinetic energy of motion at the bottom. In the absence of friction, such systems are said to be conservative because at any given moment the total amount of energy (kinetic plus potential) remains equal to the initial work done to set the system in motion.

As the science of physics expanded to cover an ever-wider range of phenomena, it became necessary to include additional forms of energy in order to keep the total amount of energy constant for all closed systems (or to account for changes in total energy for open systems). For example, if work is done to accelerate charged particles, then some of the resultant energy will be stored in the form of electromagnetic fields and carried away from the system as radiation. In turn the electromagnetic energy can be picked up by a remote receiver (antenna) and converted back into an equivalent amount of work. With his theory of special relativity, Albert Einstein realized that energy (E) can also be stored as mass (m) and converted back into energy, as expressed by his famous equation $E = mc2$, where c is the velocity of light. All of these systems are said to be conservative in the sense that energy can be freely converted from one form to another without limit. Each fundamental advance of physics into new realms has involved a similar extension to the list of the different forms of energy. In addition to preserving the first law of thermodynamics (see below), also called the law of conservation of energy, each form of energy can be related back to an equivalent amount of work required to set the system into motion.

Thermodynamics encompasses all of these forms of energy, with the further addition of heat to the list of different kinds of energy. However, heat is fundamentally different from the others in that the conversion of work (or other forms of energy) into heat is not completely reversible, even in principle. In the example of the rolling car, some of the work done to set the car in motion is inevitably lost as heat due to friction, and the car eventually comes to a stop on a level surface. Even if all the generated heat were collected and stored in some fashion, it could never be converted entirely back into mechanical energy of motion. This fundamental limitation is expressed quantitatively by the second law of thermodynamics (see below).

The role of friction in degrading the energy of mechanical systems may seem simple and obvious, but the quantitative connection between heat and work, as first discovered by Count Rumford, played a key role in understanding the operation of steam engines in the 19th century and similarly for all energy-conversion processes today.

Total internal energy

Although classical thermodynamics deals exclusively with the macroscopic properties of materials—such as temperature, pressure, and volume—thermal energy from the addition of heat can be understood at the microscopic level as an increase in the kinetic energy of motion of the molecules making up a substance. For example, gas molecules have translational kinetic energy that is proportional to the temperature of the gas: the molecules can rotate about their centre of mass, and the constituent atoms can vibrate with respect to each other (like masses connected by springs). Additionally, chemical energy is stored in the bonds holding the molecules together, and weaker long-range interactions between the molecules involve yet more energy. The sum total of all these forms of energy constitutes the total internal energy of the substance in a given thermodynamic state. The total energy of a system includes its internal energy plus any other forms of energy, such as kinetic energy due to motion of the system as a whole (e.g., water flowing through a pipe) and gravitational potential energy due to its elevation.

The sweeping generality of the constraints imposed by the laws of thermodynamics makes the number of potential applications so large that it is impractical to catalog every possible formula that might come into use, even in detailed textbooks on the subject. For this reason, students and practitioners in the field must be proficient in mathematical manipulations involving partial derivatives and in understanding their physical content.

One of the great strengths of classical thermodynamics is that the predictions for the direction of spontaneous change are completely independent of the microscopic structure of matter, but this also represents a limitation in that no predictions are made about the rate at which a system approaches equilibrium. In fact, the rate can be exceedingly slow, such as the spontaneous transition of diamonds into graphite. Statistical thermodynamics provides information on the rates of processes, as well as important insights into the statistical nature of entropy and the second law of thermodynamics.

The 20th-century English scientist C.P. Snow explained the first three laws of thermodynamics, respectively, as:

1. You cannot win (i.e., one cannot get something for nothing, because of the conservation of matter and energy).
2. You cannot break even (i.e., one cannot return to the same energy state, because entropy, or disorder, always increases).
3. You cannot get out of the game (i.e., absolute zero is unattainable because no perfectly pure substance exists).

12

Nuclear Power

---•ρ•---

Nuclear power, electricity generated by power plants that derive their heat from fission in a nuclear reactor. Except for the reactor, which plays the role of a boiler in a fossil-fuel power plant, a nuclear power plant is similar to a large coal-fired power plant, with pumps, valves, steam generators, turbines, electric generators, condensers, and associated equipment.

World nuclear power

Nuclear power provides almost 15 percent of the world's electricity. The first nuclear power plants, which were small demonstration facilities, were built in the 1960s. These prototypes provided "proof-of-concept" and laid the groundwork for the development of the higher-power reactors that followed.

The nuclear power industry went through a period of remarkable growth until about 1990, when the portion of electricity generated by nuclear power reached a high of 17 percent. That percentage remained stable through the 1990s and began to decline slowly around the turn of the 21st century, primarily because of the fact that total electricity generation grew faster than electricity from nuclear power while other sources of energy (particularly coal and natural gas) were able to grow more quickly to meet the rising demand. This trend appears likely to continue well into the 21st century. The Energy Information Administration (EIA), a statistical arm of the U.S. Department of Energy, has projected that world electricity generation between 2005 and 2035 will roughly double (from more than 15,000 terawatt-hours to 35,000 terawatt-hours) and that generation from all energy sources except petroleum will continue to grow.

In 2012 more than 400 nuclear reactors were in operation in 30 countries around the world, and more than 60 were under construction. The United

States has the largest nuclear power industry, with more than 100 reactors; it is followed by France, which has more than 50. Of the top 15 electricity-producing countries in the world, all but two, Italy and Australia, utilize nuclear power to generate some of their electricity. The overwhelming majority of nuclear reactor generating capacity is concentrated in North America, Europe, and Asia. The early period of the nuclear power industry was dominated by North America (the United States and Canada), but in the 1980s that lead was overtaken by Europe. The EIA projects that Asia will have the largest nuclear capacity by 2035, mainly because of an ambitious building program in China.

A typical nuclear power plant has a generating capacity of approximately one gigawatt (GW; one billion watts) of electricity. At this capacity, a power plant that operates about 90 percent of the time (the U.S. industry average) will generate about eight terawatt-hours of electricity per year. The predominant types of power reactors are pressurized water reactors (PWRs) and boiling water reactors (BWRs), both of which are categorized as light water reactors (LWRs) because they use ordinary (light) water as a moderator and coolant. LWRs make up more than 80 percent of the world's nuclear reactors, and more than three-quarters of the LWRs are PWRs.

Issues affecting nuclear power

Countries may have a number of motives for deploying nuclear power plants, including a lack of indigenous energy resources, a desire for energy independence, and a goal to limit greenhouse gas emissions by using a carbon-free source of electricity. The benefits of applying nuclear power to these needs are substantial, but they are tempered by a number of issues that need to be considered, including the safety of nuclear reactors, their cost, the disposal of radioactive waste, and a potential for the nuclear fuel cycle to be diverted to the development of nuclear weapons. All of these concerns are discussed below.

Safety

The safety of nuclear reactors has become paramount since the Fukushima accident of 2011. The lessons learned from that disaster included the need to (1) adopt risk-informed regulation, (2) strengthen management systems so that decisions made in the event of a severe accident are based on safety and not cost or political repercussions, (3) periodically assess new information on risks posed by natural hazards such as earthquakes and associated tsunamis, and (4) take steps to mitigate the possible consequences of a station blackout.

The four reactors involved in the Fukushima accident were first-generation BWRs designed in the 1960s. Newer Generation III designs, on the other hand, incorporate improved safety systems and rely more on so-called passive safety designs (i.e., directing cooling water by gravity rather than moving it by pumps) in order to keep the plants safe in the event of a severe accident or station blackout. For instance, in the Westinghouse AP1000 design, residual heat would be removed from the reactor by water circulating under the influence of gravity from reservoirs located inside the reactor's containment structure. Active and passive safety systems are incorporated into the European Pressurized Water Reactor (EPR) as well.

Traditionally, enhanced safety systems have resulted in higher construction costs, but passive safety designs, by requiring the installation of far fewer pumps, valves, and associated piping, may actually yield a cost saving.

Economics

A convenient economic measure used in the power industry is known as the levelized cost of electricity, or LCOE, which is the cost of generating one kilowatt-hour (kWh) of electricity averaged over the lifetime of the power plant. The LCOE is also known as the "busbar cost," as it represents the cost of the electricity up to the power plant's busbar, a conducting apparatus that links the plant's generators and other components to the distribution and transmission equipment that delivers the electricity to the consumer.

The busbar cost of a power plant is determined by 1) capital costs of construction, including finance costs, 2) fuel costs, 3) operation and maintenance (O&M) costs, and 4) decommissioning and waste-disposal costs. For nuclear power plants, busbar costs are dominated by capital costs, which can make up more than 70 percent of the LCOE. Fuel costs, on the other hand, are a relatively small factor in a nuclear plant's LCOE (less than 20 percent). As a result the cost of electricity from a nuclear plant is very sensitive to construction costs and interest rates but relatively insensitive to the price of uranium. Indeed, the fuel costs for coal-fired plants tend to be substantially greater than those for nuclear plants. Even though fuel for a nuclear reactor has to be fabricated, the cost of nuclear fuel is substantially less than the cost of fossil fuel per kilowatt-hour of electricity generated. This fuel cost advantage is due to the enormous energy content of each unit of nuclear fuel compared to fossil fuel.

The O&M costs for nuclear plants tend to be higher than those for fossil-fuel plants because of the complexity of a nuclear plant and the regulatory issues that arise during the plant's operation. Costs for decommissioning and waste disposal are included in fees charged by electrical utilities. In the United States, nuclear-generated electricity was assessed a fee of $0.001 per kilowatt-hour to pay for a permanent repository of high-level nuclear waste. This seemingly modest fee yielded about $750 million per year for the Nuclear Waste Fund.

At the beginning of the 21st century, electricity from nuclear plants typically cost less than electricity from coal-fired plants, but this formula may not apply to the newer generation of nuclear power plants, given the sensitivity of busbar costs to construction costs and interest rates. Another major uncertainty is the possibility of carbon taxes or stricter regulations on carbon dioxide emissions. These measures would almost certainly raise the operating costs of coal plants and thus make nuclear power more competitive.

Radioactive-waste disposal

Spent nuclear reactor fuel and the waste stream generated by fuel reprocessing contain radioactive materials and must be conditioned for permanent disposal. The amount of waste coming out of the nuclear fuel cycle is very small compared with the amount of waste generated by fossil fuel plants. However, nuclear waste is highly radioactive (hence its designation as high-level waste, or HLW), which makes it very dangerous to the public and the environment. Extreme care must be taken to ensure that it is stored safely and securely, preferably deep underground in permanent geologic repositories.

Despite years of research into the science and technology of geologic disposal, no permanent disposal site is in use anywhere in the world. In the last decades of the 20th century, the United States made preparations for constructing a repository for commercial HLW beneath Yucca Mountain, Nevada, but by the turn of the 21st century, this facility had been delayed by legal challenges and political decisions. Pending construction of a long-term repository, U.S. utilities have been storing HLW in so-called dry casks aboveground. Some other countries using nuclear power, such as Finland, Sweden, and France, have made more progress and expect to have HLW repositories operational in the period 2020–25.

Proliferation

The claim has long been made that the development and expansion of commercial nuclear power led to nuclear weapons proliferation, because elements of the nuclear fuel cycle (including uranium enrichment and spent-fuel reprocessing) can also serve as pathways to weapons development. However, the history of nuclear weapons development does not support the notion of a necessary connection between weapons proliferation and commercial nuclear power.

The first pathway to proliferation, uranium enrichment, can lead to a nuclear weapon based on highly enriched uranium (see nuclear weapon: Principles of atomic (fission) weapons). It is considered relatively straightforward for a country to fabricate a weapon with highly enriched uranium, but the impediment historically has been the difficulty of the enrichment process. Since nuclear reactor fuel for LWRs is only slightly enriched (less than 5 percent of the fissile isotope uranium-235) and weapons need a minimum of 20 percent enriched uranium, commercial nuclear power is not a viable pathway to obtaining highly enriched uranium.

The second pathway to proliferation, reprocessing, results in the separation of plutonium from the highly radioactive spent fuel. The plutonium can then be used in a nuclear weapon. However, reprocessing is heavily guarded in those countries where it is conducted, making commercial reprocessing an unlikely pathway for proliferation. Also, it is considered more difficult to construct a weapon with plutonium versus highly enriched uranium.

More than 20 countries have developed nuclear power industries without building nuclear weapons. On the other hand, countries that have built and tested nuclear weapons have followed other paths than purchasing commercial nuclear reactors, reprocessing the spent fuel, and obtaining plutonium. Some have built facilities for the express purpose of enriching uranium; some have built plutonium production reactors; and some have surreptitiously diverted research reactors to the production of plutonium. All these pathways to nuclear proliferation have been more effective, less expensive, and easier to hide from prying eyes than the commercial nuclear power route. Nevertheless, nuclear proliferation remains a highly sensitive issue, and any country that wishes to launch a commercial nuclear power industry will necessarily draw the close attention of oversight bodies such as the International Atomic Energy Agency.

13

Emissions Trading

----•◊•----

Emissions trading, an environmental policy that seeks to reduce air pollution efficiently by putting a limit on emissions, giving polluters a certain number of allowances consistent with those limits, and then permitting the polluters to buy and sell the allowances. The trading of a finite number of allowances results in a market price being put on emissions, which enables polluters to work out the most cost-effective means of reaching the required reduction. Emissions trading has been used with notable success to reduce emissions that cause acid rain, and it is currently being used in various attempts around the world to control emissions of greenhouse gases.

Emissions trading in principle

An idealized trading scheme might work in the following manner: A regulating authority might assign polluters a certain number of allowances defining the amount of pollutants they are permitted to emit that year. The total number of allowances would represent a certain reduction over the year before, and they would probably be scheduled to go down each subsequent year in order to reach the long-term reduction targets. One group of polluters might be able to take action during the year at relatively little cost that would actually reduce their emissions well below their allowances. In that case, they would face the prospect of finishing the year with unused allowances. A second group of polluters, meanwhile, might find it very expensive to reach their own reduction goals. In order to avoid this cost but also to avoid being fined by the regulating authority for exceeding their allowances, the second group of polluters might be willing to buy unused allowances from the first group—in effect, paying the first group to undertake the extra reductions that are too expensive for the second group.

The two would then negotiate a price for the allowances, and the agreed-upon reductions would be undertaken.

The regulating authority would not be concerned with who owned the unused allowances, so long as the total emissions were reduced. Over time, as emissions limits were progressively lowered, the allowances would become fewer in number and fetch a higher price on the market. At some point even the most severe polluter might find it cheaper to invest in pollution reduction than to purchase expensive allowances, though this would not necessarily be the case; some polluters might continue to emit above their allowed levels indefinitely, so long as other polluters were still able to sell them unused allowances at an affordable price. Polluters would continue to invest in emissions-reduction schemes or in emissions trading, depending on which was less expensive at any given time, until the overall reduction target was met.

Acid rain and greenhouse gases

The economic principles behind trading in emissions were explained by American economist Thomas Crocker in his 1966 essay "The Structuring of Atmospheric Pollution Control Systems" and by Canadian economist John H. Dales in his landmark book Pollution, Property, and Prices: An Essay in Policy-Making and Economics (1968). Emissions trading received its first large-scale practical application in the Acid Rain Program run by the U.S. Environmental Protection Agency in the 1990s. In 1990, amendments to the U.S. Clean Air Act of 1970 called for a halving of emissions of sulfur dioxide (SO_2) within two decades, along with a parallel ambitious reduction in emissions of nitrogen oxides. Emissions of SO_2, mainly by electric power plants, were eventually to be "capped" at 8.95 million tons per year in the continental United States—as opposed to the approximately 17 million tons emitted in 1980. Beginning in 1995, a growing number of power plants (eventually reaching more than 1,000) were brought into the program. Each plant was given a number of annual emission allowances consistent with the nationwide cap, and each plant's management was left to its own devices either to align its actual emissions with its allowances or to buy allowances from plants that had reduced their emissions below their yearly allowances. By 2010, power plants included in the Acid Rain Program were emitting about five million tons of SO_2 per year—well below the program's cap—and North America's acid rain problem was universally considered to have been brought under control. Industry and government officials agreed that the reductions were accomplished more efficiently under the cap-and-trade program than they would have been under a more traditional "command-and-control" system of regulations that would have specified how, when, and by how much at each plant emissions were to be reduced.

The world's first multilateral trading scheme for greenhouse gas emissions was the European Union Emissions Trading Scheme (EU ETS), established in 2005 in response to goals set by the Kyoto Protocol of 1997. The EU ETS is a cap-and-trade system similar in theory to the U.S. Acid Rain Program but vastly more complicated in practice, covering more than 10,000 large installations, from power plants to iron and steel mills as well as all of transport, including flights of non-EU airlines that arrive and depart from EU airports. Among other ambitious goals, the EU ETS aims to reduce the EU's emissions of greenhouse gases (particularly carbon dioxide) to 20 percent below 1990 levels by the year 2020.

Other so-called carbon-trading schemes exist outside the EU, though none is as ambitious or as complex. Some are limited to individual regions (e.g., Alberta, California), some are undertaken by a collection of regional governments (e.g., the Regional Greenhouse Gas Initiative in the northeastern United States), and some are organized countrywide (e.g., New Zealand, Australia).

Some proponents of emissions trading argue that no system will be truly effective at reducing greenhouse gases until it is joined by all the world's major emitters, including not just the EU but also the United States, China, and India. The linking of emissions-trading schemes around the world under the umbrella of internationally agreed-upon reduction targets, so their argument goes, would create a global price on carbon, and a globally accepted price on carbon would in turn eventually result in an efficient reduction of greenhouse gases. Some other analysts, however, argue that no emissions-trading scheme could efficiently reduce greenhouse gases, especially on a global scale. First, they argue, the damage caused to the global environment by each incremental emission of CO_2 is very small and perhaps unknowable, making it very hard to put an accurate price on emissions. Second, a global cap-and-trade system would be very difficult to administer and almost impossible to enforce. Political opponents of emissions trading add the argument that any cap-and-trade arrangement would be an unnecessary and burdensome tax on economic activity.

14

Compostable Plastics

---◦▷◦---

Compostable plastics are the next generation of plastics- they come from renewable materials and break down through composting.

Instead of using plastic made from petrochemicals and fossil fuels, compostable plastics are derived from renewable materials like corn, potato, and tapioca starches, cellulose, soy protein, and lactic acid. Compostable plastics are non-toxic and decompose back into carbon dioxide, water, and biomass when composted.

Don't get confused- compostable plastics are not the same as biodegradable, oxo-biodegradable, or bio-based conventional plastics. Some of the first alternative plastics were hybrid plastics made of both petroleum-based and plant-based resins. These hybrid plastics were not truly compostable because they contained petroleum.

What is compostable plastic made out of?

World Centric® compostable plastics are made from Ingeo™, a resin made from polylactic acid (PLA). Ingeo is made from dextrose, a sugar produced by plants. Currently, the most common raw material for Ingeo is field corn, although other plant sources may be used in the future. On average, the production of PLA resin uses about 52% less energy than the production of petroleum-based resins. Similarly, manufacturing PLA resin produces 80% less greenhouse gases than traditional petroleum-based resin

15
Supernova

———•♄•———

Supernova, plural supernovae or supernovas, any of a class of violently exploding stars whose luminosity after eruption suddenly increases many millions of times its normal level.

The term supernova is derived from nova (Latin: "new"), the name for another type of exploding star. Supernovae resemble novae in several respects. Both are characterized by a tremendous, rapid brightening lasting for a few weeks, followed by a slow dimming. Spectroscopically, they show blue-shifted emission lines, which imply that hot gases are blown outward. But a supernova explosion, unlike a nova outburst, is a cataclysmic event for a star, one that essentially ends its active (i.e., energy-generating) lifetime. When a star "goes supernova," considerable amounts of its matter, equaling the material of several Suns, may be blasted into space with such a burst of energy as to enable the exploding star to outshine its entire home galaxy.

Supernovae explosions release not only tremendous amounts of radio waves and X-rays but also cosmic rays. Some gamma-ray bursts have been associated with supernovae. Supernovae also release many of the heavier elements that make up the components of the solar system, including Earth, into the interstellar medium. Spectral analyses show that abundances of the heavier elements are greater than normal, indicating that these elements do indeed form during the course of the explosion. The shell of a supernova remnant continues to expand until, at a very advanced stage, it dissolves into the interstellar medium.

Historical supernovae

Historically, only seven supernovae are known to have been recorded before the early 17th century. The most famous of them occurred in 1054 and was seen in one of the horns of the constellation Taurus. The remnants of this explosion are visible today as the Crab Nebula, which is composed of glowing ejecta of gases flying outward in an irregular fashion and a rapidly spinning, pulsating neutron star, called a pulsar, in the centre. The supernova of 1054 was recorded by Chinese and Korean observers; it also may have been seen by southwestern American Indians, as suggested by certain rock paintings discovered in Arizona and New Mexico. It was bright enough to be seen during the day, and its great luminosity lasted for weeks. Other prominent supernovae are known to have been observed from Earth in 185, 393, 1006, 1181, 1572, and 1604.

The closest and most easily observed of the hundreds of supernovae that have been recorded since 1604 was first sighted on the morning of Feb. 24, 1987, by the Canadian astronomer Ian K. Shelton while working at the Las Campanas Observatory in Chile. Designated SN 1987A, this formerly extremely faint object attained a magnitude of 4.5 within just a few hours, thus becoming visible to the unaided eye. The newly appearing supernova was located in the Large Magellanic Cloud at a distance of about 160,000 light-years. It immediately became the subject of intense observation by astronomers throughout the Southern Hemisphere and was observed by the Hubble Space Telescope. SN 1987A's brightness peaked in May 1987, with a magnitude of about 2.9, and slowly declined in the following months.

Types of supernovae

Supernovae may be divided into two broad classes, Type I and Type II, according to the way in which they detonate. Type I supernovae may be up to three times brighter than Type II; they also differ from Type II supernovae in that their spectra contain no hydrogen lines and they expand about twice as rapidly.

Type II supernovae

The so-called classic explosion, associated with Type II supernovae, has as progenitor a very massive star (a Population I star) of at least eight solar masses that is at the end of its active lifetime. (These are seen only in spiral galaxies, most often near the arms.) Until this stage of its evolution, the star has shone by means of the nuclear energy released at and near its core in the process of squeezing and heating lighter elements such as hydrogen or helium into successively heavier elements—i.e., in the process of nuclear fusion. Forming elements heavier than iron absorbs rather than produces energy, however, and, since energy is no longer available, an iron core is built up at the centre of the aging, heavyweight star. When the iron core becomes too massive, its ability to support itself by means of the outward explosive thrust of internal fusion reactions fails to counteract the tremendous pull of its own gravity. Consequently, the core collapses. If the core's mass is less than about three solar masses, the collapse continues until the core reaches a point at which its constituent nuclei and free electrons are crushed together into a hard, rapidly spinning core. This core consists almost entirely of neutrons, which are compressed in a volume only 20 km (12 miles) across but whose combined weight equals that of several Suns. A teaspoonful of this extraordinarily dense material would weigh 50 billion tons on Earth. Such an object is called a neutron star.

The supernova detonation occurs when material falls in from the outer layers of the star and then rebounds off the core, which has stopped collapsing and suddenly presents a hard surface to the infalling gases. The shock wave generated by this collision propagates outward and blows off the star's outer gaseous layers. The amount of material blasted outward depends on the star's original mass.

If the core mass exceeds three solar masses, the core collapse is too great to produce a neutron star; the imploding star is compressed into an even smaller and denser body—namely, a black hole. Infalling material disappears into the black hole, the gravitational field of which is so intense that not even light can escape. The entire star is not taken in by the black hole, since much of the falling envelope of the star either rebounds from the temporary formation of a spinning neutron core or misses passing through the very centre of the core and is spun off instead.

Type I supernovae

Type I supernovae can be divided into three subgroups—Ia, Ib, and Ic—on the basis of their spectra. The exact nature of the explosion mechanism in Type I generally is still uncertain, although Ia supernovae, at least, are thought to originate in binary systems consisting of a moderately massive star and a white dwarf, with material flowing to the white dwarf from its larger companion. A thermonuclear explosion results if the flow of material is sufficient to raise the mass of the white dwarf above the Chandrasekhar limit of 1.44 solar masses. Unlike the case of an ordinary nova, for which the mass flow is less and only a superficial explosion results, the white dwarf in a Ia supernova explosion is presumably destroyed completely. Radioactive elements, notably nickel-56, are formed. When nickel-56 decays to cobalt-56 and the latter to iron-56, significant amounts of energy are released, providing perhaps most of the light emitted during the weeks following the explosion.

Type Ia supernovae are useful probes of the structure of the universe, since they all have the same luminosity. By measuring the apparent brightness of these objects, one also measures the expansion rate of the universe and that rate's variation with time. Dark energy, a repulsive force that is the dominant component (73 percent) of the universe, was discovered in 1998 with this method. Type Ia supernovae that exploded when the universe was only two-thirds of its present size were fainter and thus farther away than they would be in a universe without dark energy. This implies that the expansion rate of the universe is faster now than it was in the past, a result of the current dominance of dark energy. (Dark energy was negligible in the early universe.)

16

Asteroids

———•♡•———

Asteroid, also called minor planet or planetoid, any of a host of small bodies, about 1,000 km (600 miles) or less in diameter, that orbit the Sun primarily between the orbits of Mars and Jupiter in a nearly flat ring called the asteroid belt. It is because of their small size and large numbers relative to the major planets that asteroids are also called minor planets. The two designations have been used interchangeably, though the term asteroid is more widely recognized by the general public. Among scientists, those who study individual objects with dynamically interesting orbits or groups of objects with similar orbital characteristics generally use the term minor planet, whereas those who study the physical properties of such objects usually refer to them as asteroids. The distinction between asteroids and meteoroids having the same origin is culturally imposed and is basically one of size. Asteroids that are approximately house-sized (a few tens of metres across) and smaller are often called meteoroids, though the choice may depend somewhat on context—for example, whether they are considered objects orbiting in space (asteroids) or objects having the potential to collide with a planet, natural satellite, or other comparatively large body or with a spacecraft (meteoroids).

Major milestones in asteroid research

Early discoveries

The first asteroid was discovered on January 1, 1801, by the astronomer Giuseppe Piazzi at Palermo, Italy. At first Piazzi thought he had discovered a comet; however, after the orbital elements of the object had been computed, it became clear that the object moved in a planetlike orbit between the orbits of Mars and Jupiter. Because of illness, Piazzi was able to observe the object only until February 11. Although the discovery was reported in the press,

Piazzi only shared details of his observations with a few astronomers and did not publish a complete set of his observations until months later. With the mathematics then available, the short arc of observations did not allow computation of an orbit of sufficient accuracy to predict where the object would reappear when it moved back into the night sky, so some astronomers did not believe in the discovery at all.

There matters might have stood had it not been for the fact that that object was located at the heliocentric distance predicted by Bode's law of planetary distances, proposed in 1766 by the German astronomer Johann D. Titius and popularized by his compatriot Johann E. Bode, who used the scheme to advance the notion of a "missing" planet between Mars and Jupiter. The discovery of the planet Uranus in 1781 by the British astronomer William Herschel at a distance that closely fit the distance predicted by Bode's law was taken as strong evidence of its correctness. Some astronomers were so convinced that they agreed during an astronomical conference in 1800 to undertake a systematic search. Ironically, Piazzi was not a party to that attempt to locate the missing planet. Nonetheless, Bode and others, on the basis of the preliminary orbit, believed that Piazzi had found and then lost it. That led German mathematician Carl Friedrich Gauss to develop in 1801 a method for computing the orbit of minor planets from only a few observations, a technique that has not been significantly improved since. The orbital elements computed by Gauss showed that, indeed, the object moved in a planetlike orbit between the orbits of Mars and Jupiter. Using Gauss's predictions, German Hungarian astronomer Franz von Zach (ironically, the one who had proposed making a systematic search for the "missing" planet) rediscovered Piazzi's object on December 7, 1801. (It was also rediscovered independently by German astronomer Wilhelm Olbers on January 2, 1802.) Piazzi named that object Ceres after the ancient Roman grain goddess and patron goddess of Sicily, thereby initiating a tradition that continues to the present day: asteroids are named by their discoverers (in contrast to comets, which are named for their discoverers).

The discovery of three more faint objects in similar orbits over the next six years—Pallas, Juno, and Vesta—complicated that elegant solution to the missing-planet problem and gave rise to the surprisingly long-lived though no longer accepted idea that the asteroids were remnants of a planet that had exploded.

Following that flurry of activity, the search for the planet appears to have been abandoned until 1830, when Karl L. Hencke renewed it. In 1845 he discovered a fifth asteroid, which he named Astraea.

The name asteroid (Greek for "starlike") had been suggested to Herschel by classicist Charles Burney, Jr., via his father, music historian Charles Burney, Sr., who was a close friend of Herschel's. Herschel proposed the term in 1802 at a meeting of the Royal Society. However, it was not accepted until the mid-19th century, when it became clear that Ceres and the other asteroids were not planets.

There were 88 known asteroids by 1866, when the next major discovery was made: Daniel Kirkwood, an American astronomer, noted that there were gaps (now known as Kirkwood gaps) in the distribution of asteroid distances from the Sun (see below Distribution and Kirkwood gaps). The introduction of photography to the search for new asteroids in 1891, by which time 322 asteroids had been identified, accelerated the discovery rate. The asteroid designated (323) Brucia, detected in 1891, was the first to be discovered by means of photography. By the end of the 19th century, 464 had been found, and that number grew to 108,066 by the end of the 20th century and was more than 750,000 in the second decade of the 21st century. The explosive growth was a spin-off of a survey designed to find 90 percent of asteroids with diameters greater than 1 km that can cross Earth's orbit and thus have the potential to collide with the planet (see below Near-Earth asteroids).

Later advances

In 1918 the Japanese astronomer Hirayama Kiyotsugu recognized clustering in three of the orbital elements (semimajor axis, eccentricity, and inclination) of various asteroids. He speculated that objects sharing those elements had been formed by explosions of larger parent asteroids, and he called such groups of asteroids "families."

In the mid-20th century, astronomers began to consider the idea that, during the formation of the solar system, Jupiter was responsible for interrupting the accretion of a planet from a swarm of planetesimals located about 2.8 astronomical units (AU) from the Sun; for elaboration of this idea, see below Origin and evolution of the asteroids. (One astronomical unit is the average distance from Earth to the Sun—about 150 million km [93 million miles].) About the same time, calculations of the lifetimes of asteroids whose orbits passed close to those of the major planets showed that most such asteroids were destined either to collide with a planet or to be ejected from the solar system on timescales of a few hundred thousand to a few million years. Since the age of the solar system is approximately

4.6 billion years, this meant that the asteroids seen today in such orbits must have entered them recently and implied that there was a source for those asteroids. At first that source was thought to be comets that had been captured by the planets and that had lost their volatile material through repeated passages inside the orbit of Mars. It is now known that most such objects come from regions in the main asteroid belt near Kirkwood gaps and other orbital resonances.

During much of the 19th century, most discoveries concerning asteroids were based on studies of their orbits. The vast majority of knowledge about the physical characteristics of asteroids—for example, their size, shape, rotation period, composition, mass, and density—was learned beginning in the 20th century, in particular since the 1970s. As a result of such studies, those objects went from being merely "minor" planets to becoming small worlds in their own right. The discussion below follows that progression in knowledge, focusing first on asteroids as orbiting bodies and then on their physical nature.

Geography of the asteroid belt

Geography in its most-literal sense is a description of the features on the surface of Earth or another planet. Three coordinates—latitude, longitude, and altitude—suffice for locating all such features. Similarly, the location of any object in the solar system can be specified by three parameters—heliocentric ecliptic longitude, heliocentric ecliptic latitude, and heliocentric distance. Such positions, however, are valid for only an instant of time, since all objects in the solar system are continuously in motion. Thus, a better descriptor of the "location" of a solar system object is the path, called the orbit, that it follows around the Sun (or, in the case of a planetary satellite [moon], the path around its parent planet).

All asteroids orbit the Sun in elliptical orbits and move in the same direction as the major planets. Some elliptical orbits are very nearly circles, whereas others are highly elongated (eccentric). An orbit is completely described by six geometric parameters called its elements. Orbital elements, and hence the shape and orientation of the orbit, also vary with time because each object is gravitationally acting on, and being acted upon by, all other bodies in the solar system. In most cases such gravitational effects can be accounted for so that accurate predictions of past and future locations can be made and a mean orbit can be defined. Those mean orbits can then be used to describe the geography of the asteroid belt.

Names and orbits of asteroids

Because of their widespread occurrence, asteroids are assigned numbers as well as names. The numbers are assigned consecutively after accurate orbital elements have been determined. Ceres is officially known as (1) Ceres, Pallas as (2) Pallas, and so forth. Of the 757,626 asteroids discovered through 2018, 68 percent were numbered. Asteroid discoverers have the right to choose names for their discoveries as soon as they have been numbered. The names selected are submitted to the International Astronomical Union (IAU) for approval. (In 2006 the IAU decided that Ceres, the largest known asteroid, also qualified as a member of a new category of solar system objects called dwarf planets.)

Prior to the mid-20th century, asteroids were sometimes assigned numbers before accurate orbital elements had been determined, so some numbered asteroids could not later be located. Such objects were referred to as "lost" asteroids. The final lost numbered asteroid, (719) Albert, was recovered in 2000 after a lapse of 89 years. Many newly discovered asteroids still become "lost" because of an insufficiently long span of observations, but no new asteroids are assigned numbers until their orbits are reliably known.

The Minor Planet Center at the Harvard-Smithsonian Center for Astrophysics in Cambridge, Massachusetts, maintains computer files for all measurements of asteroid positions. As of 2018, there were more than 188 million such positions in its database.

Distribution and Kirkwood gaps

The great majority of the known asteroids move in orbits between those of Mars and Jupiter. Most of those orbits, in turn, have semimajor axes, or mean distances from the Sun, between 2.06 and 3.28 AU, a region called the main belt. The mean distances are not uniformly distributed but exhibit population depletions, or "gaps." Those so-called Kirkwood gaps are due to mean-motion resonances with Jupiter's orbital period. An asteroid with a mean distance from the Sun of 2.50 AU, for example, makes three circuits around the Sun in the time it takes Jupiter, which has a mean distance of 5.20 AU, to make one circuit. The asteroid is thus said to be in a three-to-one (written 3:1) resonance orbit with Jupiter. Consequently, once every three orbits, Jupiter and an asteroid in such an orbit would be in the same relative positions, and the asteroid would experience a gravitational force in a fixed direction. Repeated applications of that force would eventually change the mean distance of that asteroid—and others in similar orbits—thus creating a gap at 2.50 AU. Major gaps occur at distances from the Sun that correspond to resonances with Jupiter of 4:1, 3:1, 5:2, 7:3, and 2:1, with the respective mean distances being 2.06, 2.50, 2.82, 2.96, and 3.28. The major gap at the 4:1 resonance defines the nearest extent of the main belt; the gap at the 2:1 resonance, the farthest extent.

Some mean-motion resonances, rather than dispersing asteroids, are observed to collect them. Outside the limits of the main belt, asteroids cluster near resonances of 5:1 (at 1.78 AU, called the Hungaria group), 7:4 (at 3.58 AU, the Cybele group), 3:2 (at 3.97 AU, the Hilda group), 4:3 (at 4.29 AU, the Thule group), and 1:1 (at 5.20 AU, the Trojan groups). (See below Hungarias and outer-belt asteroids and Trojan asteroids for additional discussion of these groups.) The presence of other resonances, called secular resonances, complicates the situation, particularly at the sunward edge of the belt. Secular resonances, in which two orbits interact through the motions of their ascending nodes, perihelia, or both, operate over timescales of millions of years to change the eccentricity and inclination of asteroids. Combinations of mean-motion and secular resonances can either result in long-term stabilization of asteroid orbits at certain mean-motion resonances, as is evidenced by the Hungaria, Cybele, Hilda, and Trojan asteroid groups, or cause the orbits to evolve away from the resonances, as is evidenced by the Kirkwood gaps.

Near-Earth asteroids

Asteroids that can come close to Earth are called near-Earth asteroids (NEAs), although not all NEAs actually cross Earth's orbit. NEAs are divided into several orbital classes. Asteroids belonging to the class most distant from Earth—those asteroids that can cross the orbit of Mars but that have perihelion distances greater than 1.3 AU—are dubbed Mars crossers. That class is further subdivided into two: shallow Mars crossers (perihelion distances no less than 1.58 AU but less than 1.67 AU) and deep Mars crossers (perihelion distances greater than 1.3 AU but less than 1.58 AU).

The next-most-distant class of NEAs is the Amors. Members of that group have perihelion distances that are greater than 1.017 AU, which is Earth's aphelion distance, but no greater than 1.3 AU. Amor asteroids therefore do not at present cross Earth's orbit. Because of strong gravitational perturbations produced by their close approaches to Earth, however, the orbital elements of all Earth-approaching asteroids except the shallow Mars crossers change appreciably on timescales as short as years or decades. For that reason, about half the known Amors—including (1221) Amor, the namesake of the group—are part-time Earth crossers. Only asteroids that cross the orbits of planets—i.e., Earth-approaching asteroids and idiosyncratic objects such as (944) Hidalgo and Chiron (see below Asteroids in unusual orbits)—suffer significant changes in their orbital elements on timescales shorter than many millions of years.

There are two classes of NEAs that deeply cross Earth's orbit on an almost continuous basis. The first of those to be discovered were the Apollo asteroids, named for (1862) Apollo, which was discovered in 1932 but was lost shortly thereafter and not rediscovered until 1978. The mean distances of Apollo asteroids from the Sun are greater than or equal to 1 AU, and their perihelion distances are less than or equal to Earth's aphelion distance of 1.017 AU; thus, they cross Earth's orbit when near the closest points to the Sun in their own orbits. The other class of Earth-crossing asteroids is named Atens for (2062) Aten, which was discovered in 1976. The Aten asteroids have mean distances from the Sun that are less than 1 AU and aphelion distances that are greater than or equal to 0.983 AU, the perihelion distance of Earth; they cross Earth's orbit when near the farthest points from the Sun of their orbits.

The class of NEAs that was the last to be recognized is composed of asteroids with orbits entirely inside that of Earth. Known as Atira asteroids after (163693) Atira, they have mean distances from the Sun that are less than 1 AU and aphelion distances less than 0.983 AU; they do not cross Earth's orbit.

By 2018 the known Atira, Aten, Apollo, and Amor asteroids of all sizes numbered 17, 1,342, 8,972, and 7,618, respectively, although those numbers are steadily increasing as the asteroid survey programs progress. Most of those have been discovered since 1970, when dedicated searches for those types of asteroids were begun. Astronomers have estimated that there are roughly 15 Atiras, 45 Atens, 570 Apollos, and 270 Amors that have diameters larger than about 1 km (0.6 mile).

Because they can approach quite close to Earth, some of the best information available on asteroids has come from Earth-based radar studies of NEAs. In 1968 the Apollo asteroid (1566) Icarus became the first NEA to be observed with radar. By 2018 over 750 NEAs had been so observed. Because of continuing improvements to the radar systems themselves and to the computers used to process the data, the information provided by that technique increased dramatically beginning in the final decade of the 20th century. For example, the first images of an asteroid, (4769) Castalia, were made by using radar data obtained in 1989, more than two years before the first spacecraft flyby of an asteroid—(951) Gaspra by the Galileo spacecraft in 1991 (see below Spacecraft exploration). The observations of Castalia provided the first evidence in the solar system for a double-lobed object, interpreted to be two roughly equal-sized bodies in contact. Radar observations of (4179) Toutatis in 1992 revealed it to be several kilometres long with a peanut-shell shape; similar to Castalia, Toutatis appears predominantly to be two components in contact, one about twice as large as the other. The highest-resolution images show craters having diameters between 100 and 600 metres (roughly 300 and 2,000 feet). Radar images of (1620) Geographos obtained in 1994 were numerous enough and of sufficient quality for an animation to be made showing it rotating.

The orbital characteristics of NEAs mean that some of those objects make close approaches to Earth and occasionally collide with it. In January 1991, for example, an Apollo asteroid (or, as an alternative description, a large meteoroid) with an estimated diameter of 10 metres (33 feet) passed by Earth within less than half the distance to the Moon. Such passages are not especially unusual. On October 6, 2008, the asteroid 2008 TC3, which had a size of about 5 metres (16 feet), was discovered. It crashed in the Nubian Desert of Sudan the next day. However, because of the small sizes of NEAs and the short time they spend close enough to Earth to be seen, it is unusual for such close passages to be observed. An example of an NEA for which the lead time for observation is large is (99942) Apophis. That Aten asteroid, which has a diameter of about 375 metres (1,230 feet), is predicted to pass within 32,000 km (20,000 miles) of Earth—i.e., closer than communications satellites in geostationary orbits—on April 13, 2029; during that passage its probability of hitting Earth is thought to be near zero. The collision of a sufficiently large NEA with Earth is generally recognized to pose a great potential danger to human beings and possibly to all life on the planet. For a detailed discussion of this topic, see Earth impact hazard.

Main-belt asteroid families

Within the main belt are groups of asteroids that cluster with respect to certain mean orbital elements (semimajor axis, eccentricity, and inclination). Such groups are called families and are named for the lowest numbered asteroid in the family. Asteroid families are formed when an asteroid is disrupted in a catastrophic collision, the members of the family thus being pieces of the original asteroid. Theoretical studies indicate that catastrophic collisions between asteroids are common enough to account for the number of families observed. About 40 percent of the larger asteroids belong to such families, but as high a proportion as 90 percent of small asteroids (i.e., those about 1 km in diameter) may be family members, because each catastrophic collision produces many more small fragments than large ones and smaller asteroids are more likely to be completely disrupted.

The three largest families in the main asteroid belt are named Eos, Koronis, and Themis. Each family has been determined to be compositionally homogeneous; that is, all the members of a family appear to have the same basic chemical makeup. If the asteroids belonging to each family are considered to be fragments of a single parent body, then their parent bodies must have had diameters of 200, 90, and 300 km (124, 56, and 186 miles), respectively. The smaller families present in the main belt have not been as well studied, because their numbered members are fewer and smaller (and hence fainter when viewed telescopically). It is theorized that some of the Earth-crossing asteroids and the great majority of meteorites reaching Earth's surface are fragments produced in collisions similar to those that produced the asteroid families. For example, the asteroid Vesta, whose surface appears to be basaltic rock, is the parent body of the meteorites known as basaltic achondrite HEDs, a grouping of the related howardite, eucrite, and diogenite meteorite types. (For additional discussion of the HED meteorites and Vesta, see meteorite: Achondrites.)

Hungarias and outer-belt asteroids

Only one known concentration of asteroids, the Hungaria group, occupies the region between Mars and the inner edge of the main belt. The orbits of all the Hungarias lie outside the orbit of Mars, whose aphelion distance is 1.67 AU. Hungaria asteroids have nearly circular (low-eccentricity) orbits but large orbital inclinations to Earth's orbit and the general plane of the solar system.

Four known asteroid groups fall beyond the main belt but within or near the orbit of Jupiter, with mean distances from the Sun between about 3.28 and 5.3 AU, as mentioned above in the section Distribution and Kirkwood gaps. Collectively called outer-belt asteroids, they have orbital periods that range from more than one-half that of Jupiter to approximately Jupiter's period. Three of the outer-belt groups—the Cybeles, the Hildas, and Thule—are named after the lowest-numbered asteroid in each group. Members of the fourth group are called Trojan asteroids (see below). By 2015 there were about 1,894 Cybeles, 1,197 Hildas, 3 Thules, and 6,179 Trojans. Those groups should not be confused with asteroid families, all of which share a common parent asteroid. However, some of those groups—e.g., the

Hildas and Trojans—contain families.

Trojan asteroids

In 1772 the French mathematician and astronomer Joseph-Louis Lagrange predicted the existence and location of two groups of small bodies located near a pair of gravitationally stable points along Jupiter's orbit. Those are positions (now called Lagrangian points and designated L4 and L5) where a small body can be held, by gravitational forces, at one vertex of an equilateral triangle whose other vertices are occupied by the massive bodies of Jupiter and the Sun. Those positions, which lead (L4) and trail (L5) Jupiter by 60° in the plane of its orbit, are two of the five theoretical Lagrangian points in the solution to the circular restricted three-body problem of celestial mechanics (see celestial mechanics: The restricted three-body problem). The other three points are located along a line passing through the Sun and Jupiter. The presence of other planets, however—principally Saturn—perturbs the Sun-Jupiter-Trojan asteroid system enough to destabilize those points, and no asteroids have been found actually at them. In fact, because of that destabilization, most of Jupiter's Trojan asteroids move in orbits inclined as much as 40° from Jupiter's orbit and displaced as much as 70° from the leading and trailing positions of the true Lagrangian points.

In 1906 the first of the predicted objects, (588) Achilles, was discovered near the Lagrangian point preceding Jupiter in its orbit. Within a year two more were found: (617) Patroclus, located near the trailing Lagrangian point, and (624) Hektor, near the leading Lagrangian point. It was later decided to continue naming such asteroids after participants in the Trojan War as recounted in Homer's epic work the Iliad and, furthermore, to name those near the leading point after Greek warriors and those near the trailing point after Trojan warriors. With the exception of the two "misplaced" names already bestowed (Hektor, the lone Trojan in the Greek camp, and Patroclus, the lone Greek in the Trojan camp), that tradition has been maintained.

As of 2015, of the more than 6,200 Jupiter Trojan asteroids discovered, about two-thirds are located near the leading Lagrangian point, L4, and the remainder are near the trailing one, L5. Astronomers estimate that 1,800–2,200 of the total existing population of Jupiter's Trojans have diameters greater than 10 km (6 miles).

Since the discovery of Jupiter's orbital companions, the term Trojan has been applied to any small object occupying the equilateral Lagrangian points of other pairs of relatively massive bodies. Astronomers have searched for Trojan objects of Earth, Mars, Saturn, Uranus, and Neptune as well as of the Earth-Moon system. It was long considered doubtful whether truly stable orbits could exist near those Lagrangian points because of gravitational perturbations by the major planets. However, in 1990 an asteroid later named (5261) Eureka was discovered librating (oscillating) about the trailing Lagrangian point of Mars, and since then three others have been found, two at the trailing point and one at the leading point. Twelve Trojans of Neptune, all but three associated with the leading Lagrangian point, have been discovered since 2001. The first Earth Trojan asteroid, 2010 TK7, which librates around L4, was discovered in 2010. The first Uranus Trojan, 2011 QF99, which librates around L4, was discovered in 2011. Although Trojans of Saturn have yet to be found, objects librating about Lagrangian points of the systems formed by Saturn and its moon Tethys and Saturn and its moon Dione are known.

Asteroids in unusual orbits

Although most asteroids travel in fairly circular orbits, there are notable exceptions. In addition to the near-Earth asteroids, some objects are known to travel in orbits that extend far inside or outside the main belt. One of the most extreme is (3200) Phaethon, the first asteroid to be discovered by a spacecraft (the Infrared Astronomical Satellite in 1983). Phaethon approaches to within 0.14 AU of the Sun, well within the perihelion distance of 0.31 AU for Mercury, the innermost planet. By contrast, Phaethon's aphelion distance of 2.4 AU is in the main asteroid belt. That object is the parent body of the Geminid meteor stream, the concentration of meteoroids responsible for the annual Geminid meteor shower seen on Earth each December. Because the parent bodies of all other meteor streams identified to date are comets, Phaethon is considered by some to be a defunct comet—one that has lost its volatile materials and no longer displays the classic cometary features of a nebulous coma and a tail. Another asteroid, (944) Hidalgo, is also thought by some to be a defunct comet because of its unusual orbit. That object, discovered in 1920, travels sunward as near as 2.02 AU, which is at the inner edge of the main asteroid belt, and as far as 9.68 AU, which is just beyond the orbit of Saturn, at 9.54 AU.

In contrast to the examples of Phaethon and Hidalgo is Chiron, which, following its discovery in 1977, was classified as an asteroid, (2060) Chiron. In 1989 the object was observed to have a dusty coma surrounding it, and in 1991 the presence of cyanogen radicals was detected, a known constituent of the gas comas of comets. Chiron travels in an orbit that lies wholly exterior to the asteroid belt, having a perihelion distance of 8.43 AU (inside the orbit of Saturn) and an aphelion distance of 18.8 AU, which nearly reaches the orbit of Uranus at 19.2 AU. At the time of its discovery, Chiron was the most-distant asteroid known. Within a few years additional objects were discovered traveling among the orbits of the giant planets. It is now known that Chiron belongs to a group collectively referred to as Centaur objects, all of which have elongated orbits with perihelia outside the orbit of Jupiter and aphelia near the orbit of Uranus or Neptune. Centaurs are thought to be icy bodies—in essence, giant comet nuclei—that have been gravitationally perturbed out of the Kuiper belt beyond Neptune and presently travel mainly between the orbits of Jupiter and Neptune. All Centaurs move in chaotic planet-orbit-crossing orbits. Their orbits will evolve away from the Centaur region, and they will eventually collide with the Sun or a planet or be permanently ejected from the solar system. By 2015 there were 302 known Centaurs.

Asteroids traditionally have been distinguished from comets by characteristics based on physical differences, location in the solar system, and orbital properties. An object is classified as a comet when it displays "cometary activity"—i.e., a coma or tail (or any evidence of gas or dust coming from it). Objects in the Kuiper belt, all of which have mean distances from the Sun greater than that of Neptune, are considered to be comet nuclei. Because of their great distance from the Sun, however, they do not display the characteristic activity of comets. In addition, any object on a nonreturning orbit (a parabolic or hyperbolic orbit, rather than an elliptical one) is generally considered to be a comet.

Although such distinctions apply most of the time, they are not always sufficient to classify an individual object as an asteroid or a comet. For example, an object found to be receding from the Sun on a nonreturning orbit and displaying no cometary activity could be a comet, or it could be a planet-crossing asteroid being ejected from the solar system after a close encounter with a planet, most likely Jupiter. Again, objects on some planet-crossing orbits may have originated in either the Kuiper belt or the main asteroid belt. Unless such an object reveals itself by displaying cometary activity, there is usually no way to determine its origin and thus to classify it unequivocally. The object may have formed as an icy body but lost its volatile materials during a series of passes into the inner solar system. Its burned-out remnant of rocky material would presently have more physical characteristics in common with asteroids than with other comets.

The next group of asteroids to be recognized were the Kuiper belt objects (KBOs), or trans-Neptunian objects (TNOs), objects that have semimajor axes greater than that of the planet Neptune. Although not realized at the time, (134340) Pluto, discovered in 1930, was the first KBO. By 2015 more than 1,250 KBOs were known. The largest known TNOs are (136199) Eris and Pluto, of nearly equal size, followed by (136472) Makemake and (136108) Haumea. Those four TNOs, together with the asteroid Ceres, are known as dwarf planets. That is, they are physically as much "planets" as Mercury or Mars, but they have not "cleared their orbital region of other objects," an additional requirement for "planethood" imposed by the International Astronomical Union (IAU) in 2006.

There are the main-belt comets (also called active asteroids), the first member of which, 133P/(7968) Elst-Pizarro, was discovered in 1996 when that "asteroid's" cometlike activity was first noted. Main-belt comets have orbits in the main asteroid belt but exhibit cometary activity such as comae and tails.

Asteroids as individual worlds

The first measurements of the sizes of individual asteroids were made in the last years of the 19th century. A filar micrometer, an instrument normally used in conjunction with a telescope for visual measurement of the separations of double stars, was employed to estimate the diameters of the first four known asteroids. The results established that Ceres was the largest asteroid, having a diameter estimated to be nearly 800 km (500 miles). Those values remained the best available until new techniques for finding albedos (reflectivities) and diameters, based on infrared radiometry and polarization measurements, were introduced beginning about 1970 (see below Size and albedo). The first four asteroids came to be known as the "big four," and, because all other asteroids were much fainter, they all were believed to be considerably smaller as well.

The first asteroid to have its mass determined was Vesta—in 1966 from measurements of its perturbation of the orbit of asteroid (197) Arete. The first mineralogical determination of the surface composition of an asteroid was made in 1969 when spectral reflectance measurements (see below Composition) identified the mineral pyroxene in the surface material of Vesta.

Classification of asteroids

In the mid-1970s astronomers using information gathered from studies of colour, spectral reflectance, and albedo recognized that asteroids could be grouped into three broad taxonomic classes, designated C, S, and M. At that time they estimated that about 75 percent belonged to class C, 15 percent to class S, and 5 percent to class M. The remaining 5 percent were unclassifiable because of either poor data or genuinely unusual properties. Furthermore, they noted that the S class dominated the population at the inner edge of the asteroid belt, whereas the C class was dominant in the middle and outer regions of the belt.

Within a decade that taxonomic system was expanded, and it was recognized that the asteroid belt comprised overlapping rings of differing taxonomic classes, with classes designated S, C, P, and D dominating the populations at distances from the Sun of about 2, 3, 4, and 5 AU, respectively. As more data became available from further observations, additional minor classes were recognized. For discussion of the relationship of the asteroid classes to their composition,

Physical characteristics of asteroids

Rotation and shape

The rotation periods and shapes of asteroids are determined primarily by monitoring their changing brightness on timescales of minutes to days. Short-period fluctuations in brightness caused by the rotation of an irregularly shaped asteroid or a spherical spotted asteroid (i.e., one with albedo differences) produce a light curve—a graph of brightness versus time—that repeats at regular intervals corresponding to an asteroid's rotation period. The range of brightness variation is closely related to an asteroid's shape or spottedness but is more difficult to interpret.

By 2015 reliable rotation periods were known for more than 5,500 asteroids. They range from 25 seconds to 78 days, but more than two-thirds lie between 4 and 24 hours. In some cases periods longer than a few days may actually be due to precession (a smooth slow circling of the rotation axis) caused by an unseen satellite of the asteroid. Periods on the order of minutes are observed only for very small objects (those with diameters less than about 150 metres [500 feet]). The largest asteroids (those with diameters greater than about 200 km [120 miles]) have a mean rotation period close to 8 hours; the value increases to 13 hours for asteroids with diameters of about 100 km (60 miles) and then decreases to about 6 hours for those with diameters of about 10 km (6 miles). The largest asteroids may have preserved the rotation rates they had when they were formed, but the smaller ones almost certainly have had theirs modified by subsequent collisions and in the case of the very smallest, perhaps also by radiation effects. The difference in rotation periods between 200-km-class and 100-km-class asteroids is believed to stem from the fact that large asteroids retain all of the collision debris from minor collisions, whereas smaller asteroids retain more of the debris ejected in the direction opposite to that of their spins, causing a loss of angular momentum and thus a reduction in speed of rotation.

Major collisions can completely disrupt smaller asteroids. The debris from such collisions makes still smaller asteroids, which can have virtually any shape or spin rate. Thus, the fact that no rotation periods shorter than about two hours have been observed for asteroids greater than about 150 metres in diameter implies that their material strengths are not high enough to withstand the centripetal forces that such rapid spins produce.

It is impossible to distinguish mathematically between the rotation of a spotted sphere and an irregular shape of uniform reflectivity on the basis of observed brightness changes alone. Nevertheless, the fact that opposite sides of most asteroids appear to differ no more than a few percent in albedo suggests that their brightness variations are due mainly to changes in the projection of their illuminated portions as seen from Earth. Hence, in the absence of evidence to the contrary, astronomers generally accept that variations in reflectivity contribute little to the observed amplitude, or range in brightness variation, of an asteroid's rotational light curve. Vesta is a notable exception to that generalization, because the difference in reflectivity between its opposite hemispheres is known to be sufficient to account for much of its modest light-curve amplitude.

Observed light-curve amplitudes for asteroids range from zero to more than a factor of eight. There are nine reliably observed asteroids with light-curve amplitudes greater than 2.0 magnitudes; all are NEAs. They have rotation periods between 7.4 minutes and 6.8 hours and diameters between approximately 28 metres (92 feet) and 2.5 km (1.6 miles).

A rotating asteroid shows a light-curve amplitude of zero (no change in amplitude) when its shape is a uniform sphere or when it is viewed along one of its rotational axes. Before Geographos was studied by radar (see above Near-Earth asteroids), its 6.5 to 1 variation in brightness was ascribed to either of two possibilities: the asteroid is a cigar-shaped object that is being viewed along a line perpendicular to its rotational axis (which for normally rotating asteroids is the shortest axis), or it is a pair of objects nearly in contact that orbit each other around their centre of mass. The radar images ruled out the binary model, revealing that Geographos is a single highly elongated object.

The mean rotational light-curve amplitude for asteroids is a factor of about 1.3. That information, together with the assumptions discussed above, allows astronomers to estimate asteroid shapes, which occur in a wide range. Some asteroids, such as Ceres, Pallas, and Vesta, are nearly spherical, whereas others, such as (15) Eunomia, (107) Camilla, and (511) Davida, are quite elongated. Still others, as, for example, (1580) Betulia, Hektor, and Castalia (the last of which appears in radar observations to be two bodies in contact, as discussed above in Near-Earth asteroids), apparently have bizarre shapes.

Size and albedo

About 30 asteroids are larger than 200 km. The largest, Ceres, has a diameter of about 940 km (580 miles). It is followed by Vesta at 525 km (325 miles), Pallas at 510 km (320 miles), and (10) Hygiea at 410 km (250 miles). Three asteroids are between 300 and 400 km (190 and 250 miles) in diameter, and about 23 are between 200 and 300 km (120 and 190 miles). It has been estimated that 250 asteroids are larger than 100 km (60 miles) in diameter and perhaps a million are larger than 1 km (0.6 mile). The smallest known asteroids are members of the near-Earth group, some of which approach Earth to within a few hundredths of 1 AU. The smallest routinely observed Earth-approaching asteroids measure about 100 metres (330 feet) across.

The most widely used technique for determining the sizes of asteroids (and other small bodies in the solar system) is that of thermal radiometry. That technique exploits the fact that the infrared radiation (heat) emitted by an asteroid must balance the solar radiation it absorbs. By using a so-called thermal model to balance the measured intensity of infrared radiation with that of radiation at visual wavelengths, investigators are able to derive the diameter of the asteroid. Other remote-sensing techniques—for example, polarimetry, radar, and adaptive optics (techniques for minimizing the distorting effects of Earth's atmosphere)—also are used, but they are limited to brighter, larger, or closer asteroids.

The only techniques that measure the diameter directly (i.e., without having to model the actual observations) are those of stellar occultation and direct imaging using either advanced instruments on Earth (e.g., large telescopes equipped with adaptive optics or orbiting observatories such as the Hubble Space Telescope) or passing spacecraft. In the method of stellar occultation, investigators measure the length of time that a star disappears from view owing to the passage of an asteroid between the star and Earth. Then, by using the known distance and the rate of motion of the asteroid, they are able to determine the latter's diameter as projected onto the plane of the sky. For a good diameter measurement, numerous measurements across the asteroid are required, necessitating numerous observers spread out perpendicular to the asteroid's shadow track over Earth. The majority of those observations have been obtained by amateur astronomers. The necessary techniques for imaging asteroids directly were perfected during the last years of the 20th century. They (and radar) can be used to observe an asteroid over a complete rotation cycle and so measure the three-dimensional shape. Those results have made it possible to calibrate the indirect techniques, thermal radiometry in particular, such that diameter measurements made with thermal radiometry on asteroids larger than about 20 km (12 miles) are thought to be uncertain by less than 10 percent; for smaller asteroids the uncertainty is about 30 percent.

The occultation technique is limited to the relatively rare passages of asteroids in front of stars, and, because the technique measures only one cross section, it is best applied to fairly spherical asteroids. On the other hand, direct imaging (at least to date) has been limited to the nearer, brighter, or larger asteroids. Consequently, the majority of asteroid sizes have been and will probably continue to be obtained with indirect techniques. Direct imaging has allowed the accurate determination of the diameters of about two dozen asteroids, including Ceres, Pallas, Juno, and Vesta, compared with over 150,000 measured with indirect techniques, principally thermal radiometry obtained with NASA's Wide-field Infrared Survey Explorer (WISE) satellite.

A property that is closely related to size (and that also provides compositional information) is albedo. Albedo is the ratio between the amount of light actually reflected and that which would be reflected by a uniformly scattering disk of the same size, both observed at opposition. Snow has an albedo of approximately 1, and coal an albedo of about 0.05.

An asteroid's apparent brightness depends on both its albedo and its diameter as well as on its distance. For example, if Ceres and Vesta could both be observed at the same distance, Vesta would be the brighter of the two by about 15 percent, even though Vesta's diameter is only a little more than half that of Ceres. Vesta would appear brighter because its albedo is about 0.35, compared with 0.10 for Ceres.

Asteroid albedos range from about 0.02 to more than 0.5 and may be divided into four groups: low (0.02–0.07), intermediate (0.08–0.12), moderate (0.13–0.28), and high (greater than 0.28). After corrections are added for the fact that the brighter and nearer asteroids are favoured for discovery, about 78 percent of known asteroids larger than about 25 km (16 miles) in diameter are found to be low-albedo objects. Most of those are located in the outer half of the main asteroid belt and among the outer-belt populations. More than 95 percent of outer-belt asteroids belong to that group. Roughly 18 percent of known asteroids belong to the moderate-albedo group, the vast majority of which are found in the inner half of the main belt. The intermediate- and high-albedo asteroid groups make up the remaining 4 percent of the population. For the most part, they occupy the same part of the main belt as the moderate-albedo objects.

The albedo distribution for asteroids with diameters less than 25 km is poorly known because only a small fraction of that population has been characterized. However, if those objects are mostly fragments from a few asteroid families, then their albedo distribution may differ significantly from that of their larger siblings.

Mass and density

Most asteroid masses are low, although present-day observations show that the asteroids measurably perturb the orbits of the major planets. Except for Mars, however, those perturbations are too small to allow the masses of the asteroids in question to be determined. Radio-ranging measurements that were transmitted from the surface of Mars between 1976 and 1980 by the two Viking landers and time-delay radar observations using the Mars Pathfinder lander made it possible to determine distances to Mars with an accuracy of about 10 metres (33 feet). The three largest asteroids—Ceres, Vesta, and Pallas—were found to cause departures of Mars from its predicted orbit in excess of 50 metres (160 feet) over times of 10 years or less. The measured departures, in turn, were used to estimate the masses of the three asteroids. Masses for a number of other asteroids have been determined by noting their effect on the orbits of other asteroids that they approach closely and regularly, on the orbits of the asteroids' satellites, or on spacecraft orbiting or flying by the asteroids. For those asteroids whose diameters are determined and whose shapes are either spherical or ellipsoidal, their volumes are easily calculated. Knowledge of the mass and volume allows the density to be calculated. For asteroids with satellites the density can be determined directly from the satellite's orbit without knowledge of the mass.

The mass of the largest asteroid, Ceres, is 9.3 × 1020 kg, or less than 0.0002 the mass of Earth. The masses of the second and third largest asteroids, Pallas and Vesta, are each only about one-fourth the mass of Ceres. The mass of the entire asteroid belt is roughly three times that of Ceres. Most of the mass in the asteroid belt is concentrated in the larger asteroids, with about 90 percent of the total in asteroids having diameters greater than 100 km. The 10th largest asteroid has only about 1/60 the mass of Ceres. Of the total mass of the asteroids, 90 percent is located in the main belt, 9 percent is in the outer belt (including Jupiter's Trojan asteroids), and the remainder is distributed among the Hungarias and planet-crossing asteroid populations.

The densities of Ceres, Pallas, and Vesta are 2.1, 2.7, and 3.5 grams per cubic cm, respectively. Those compare with 5.4, 5.2, and 5.5 for Mercury, Venus, and Earth, respectively; 3.9 for Mars; and 3.3 for the Moon. The density of Ceres is similar to that of a class of meteorites known as carbonaceous chondrites, which contain a larger fraction of volatile material than do ordinary terrestrial rocks and hence have a somewhat lower density. The density of Pallas and Vesta are similar to those of Mars and the Moon. Insofar as Ceres, Pallas, and Vesta are typical of asteroids in general, it can be concluded that main-belt asteroids are rocky bodies.

Composition

The combination of albedos and spectral reflectance measurements—specifically, measures of the amount of reflected sunlight at wavelengths between about 0.3 and 1.1 micrometres (μm)—is used to classify asteroids into various taxonomic classes. If sufficient spectral resolution is available, especially extending to wavelengths of about 2.5 μm, those measurements also can be used to infer the composition of the surface reflecting the light. That can be done by comparing the asteroid data with data obtained in the laboratory by using meteorites or terrestrial rocks or minerals.

By the end of the 1980s, spectral reflectance measurements at wavelengths between 0.3 and 1.1 μm were available for about 1,000 asteroids, and albedos had been determined for roughly 2,000. Both types of data were available for about 400 asteroids. The table summarizes the taxonomic classes into which the asteroids are divided on the basis of such data. Starting in the 1990s, the use of detectors with improved resolution and sensitivity for spectral reflectance measurements resulted in revised taxonomies. Those versions are similar to the one presented in the table, the major difference being that the higher-resolution data have allowed many of the classes, especially the S class, to be further subdivided.

17
Water Pollution

Water pollution, the release of substances into subsurface groundwater or into lakes, streams, rivers, estuaries, and oceans to the point where the substances interfere with beneficial use of the water or with the natural functioning of ecosystems. In addition to the release of substances, such as chemicals or microorganisms, water pollution may also include the release of energy, in the form of radioactivity or heat, into bodies of water.

Sewage and other water pollutants

Water bodies can be polluted by a wide variety of substances, including pathogenic microorganisms, putrescible organic waste, plant nutrients, toxic chemicals, sediments, heat, petroleum (oil), and radioactive substances. Several types of water pollutants are considered below.

Domestic sewage

Domestic sewage is the primary source of pathogens (disease-causing microorganisms) and putrescible organic substances. Because pathogens are excreted in feces, all sewage from cities and towns is likely to contain pathogens of some type, potentially presenting a direct threat to public health. Putrescible organic matter presents a different sort of threat to water quality. As organics are decomposed naturally in the sewage by bacteria and other microorganisms, the dissolved oxygen content of the water is depleted. This endangers the quality of lakes and streams, where high levels of oxygen are required for fish and other aquatic organisms to survive. Sewage-treatment processes reduce the levels of pathogens and organics in wastewater, but they do not eliminate them completely.

Domestic sewage is also a major source of plant nutrients, mainly nitrates and phosphates. Excess nitrates and phosphates in water promote the growth of algae, sometimes causing unusually dense and rapid growths known as algal blooms. When the algae die, oxygen dissolved in the water declines because microorganisms use oxygen to digest algae during the process of decomposition then metabolize the organic wastes, releasing gases such as methane and hydrogen sulfide, which are harmful to the aerobic (oxygen-requiring) forms of life. The process by which a lake changes from a clean, clear condition—with a relatively low concentration of dissolved nutrients and a balanced aquatic community—to a nutrient-rich, algae-filled state and thence to an oxygen-deficient, waste-filled condition is called eutrophication. Eutrophication is a naturally occurring, slow, and inevitable process. However, when it is accelerated by human activity and water pollution (a phenomenon called cultural eutrophication), it can lead to the premature aging and death of a body of water.

Toxic waste

Waste is considered toxic if it is poisonous, radioactive, explosive, carcinogenic (causing cancer), mutagenic (causing damage to chromosomes), teratogenic (causing birth defects), or bioaccumulative (that is, increasing in concentration at the higher ends of food chains). Sources of toxic chemicals include improperly disposed wastewater from industrial plants and chemical process facilities (lead, mercury, chromium) as well as surface runoff containing pesticides used on agricultural areas and suburban lawns (chlordane, dieldrin, heptachlor).

Sediment

Sediment (e g., silt) resulting from soil erosion can be carried into water bodies by surface runoff. Suspended sediment interferes with the penetration of sunlight and upsets the ecological balance of a body of water. Also, it can disrupt the reproductive cycles of fish and other forms of life, and when it settles out of suspension it can smother bottom-dwelling organisms.

Thermal pollution

Heat is considered to be a water pollutant because it decreases the capacity of water to hold dissolved oxygen in solution, and it increases the rate of metabolism of fish. Valuable species of game fish (e.g., trout) cannot survive in water with very low levels of dissolved oxygen. A major source of heat is the practice of discharging cooling water from power plants into rivers; the discharged water may be as much as 15 °C (27 °F) warmer than the naturally occurring water.

Petroleum (oil) pollution

Petroleum (oil) pollution occurs when oil from roads and parking lots is carried in surface runoff into water bodies. Accidental oil spills are also a source of oil pollution—as in the devastating spills from the tanker Exxon Valdez (which released more than 260,000 barrels in Alaska's Prince William Sound in 1989) and from the Deepwater Horizon oil rig (which released more than 4 million barrels of oil into the Gulf of Mexico in 2010). Oil slicks eventually move toward shore, harming aquatic life and damaging recreation areas.

Groundwater and oceans

Groundwater—water contained in underground geologic formations called aquifers—is a source of drinking water for many people. For example, about half the people in the United States depend on groundwater for their domestic water supply. Although groundwater may appear crystal clear (due to the natural filtration that occurs as it flows slowly through layers of soil), it may still be polluted by dissolved chemicals and by bacteria and viruses. Sources of chemical contaminants include poorly designed or poorly maintained subsurface sewage-disposal systems (e.g., septic tanks), industrial wastes disposed of in improperly lined or unlined landfills or lagoons, leachates from unlined municipal refuse landfills, mining and petroleum production, and leaking underground storage tanks below gasoline service stations. In coastal areas, increasing withdrawal of groundwater (due to urbanization and industrialization) can cause saltwater intrusion: as the water table drops, seawater is drawn into wells.

Although estuaries and oceans contain vast volumes of water, their natural capacity to absorb pollutants is limited. Contamination from sewage outfall pipes, from dumping of sludge or other wastes, and from oil spills can harm marine life, especially microscopic phytoplankton that serve as food for larger aquatic organisms. Sometimes, unsightly and dangerous waste materials can be washed back to shore, littering beaches with hazardous debris. By 2010, an estimated 4.8 million and 12.7 million tonnes (between 5.3 million and 14 million tons) of plastic debris had been dumped into the oceans annually, and floating plastic waste had accummulated in Earth's five subtropical gyres that cover 40 percent of the world's oceans .

Another ocean pollution problem is the seasonal formation of "dead zones" (i.e., hypoxic areas, where dissolved oxygen levels drop so low that most higher forms of aquatic life vanish) in certain coastal areas. The cause is nutrient enrichment from dispersed agricultural runoff and concomitant algal blooms. Dead zones occur worldwide; one of the largest of these (sometimes as large as 22,730 square km [8,776 square miles]) forms annually in the Gulf of Mexico, beginning at the Mississippi River delta.

Sources of pollution

Water pollutants come from either point sources or dispersed sources. A point source is a pipe or channel, such as those used for discharge from an industrial facility or a city sewerage system. A dispersed (or nonpoint) source is a very broad, unconfined area from which a variety of pollutants enter the water body, such as the runoff from an agricultural area. Point sources of water pollution are easier to control than dispersed sources because the contaminated water has been collected and conveyed to one single point where it can be treated. Pollution from dispersed sources is difficult to control, and, despite much progress in the building of modern sewage-treatment plants, dispersed sources continue to cause a large fraction of water pollution problems.

Water quality standards

Although pure water is rarely found in nature (because of the strong tendency of water to dissolve other substances), the characterization of water quality (i.e., clean or polluted) is a function of the intended use of the water. For example, water that is clean enough for swimming and fishing may not be clean enough for drinking and cooking. Water quality standards (limits on the amount of impurities allowed in water intended for a particular use) provide a legal framework for the prevention of water pollution of all types.

There are several types of water quality standards. Stream standards are those that classify streams, rivers, and lakes on the basis of their maximum beneficial use; they set allowable levels of specific substances or qualities (e.g., dissolved oxygen, turbidity, pH) allowed in those bodies of water, based on their given classification. Effluent (water outflow) standards set specific limits on the levels of contaminants (e.g., biochemical oxygen demand, suspended solids, nitrogen) allowed in the final discharges from wastewater-treatment plants. Drinking-water standards include limits on the levels of specific contaminants allowed in potable water delivered to homes for domestic use. In the United States, the Clean Water Act and its amendments regulate water quality and set minimum standards for waste discharges for each industry as well as regulations for specific problems such as toxic chemicals and oil spills. In the European Union, water quality is governed by the Water Framework Directive, the Drinking Water Directive, and other laws.

www.ingramcontent.com/pod-product-compliance
Lightning Source LLC
Chambersburg PA
CBHW071012200526
45171CB00008B/377